Astronomers' Universe

For further volumes:
http://www.springer.com/series/6960

David S. Stevenson

Under a Crimson Sun

Prospects for Life
in a Red Dwarf System

 Springer

David S. Stevenson
Sherwood, UK

ISSN 1614-659X
ISBN 978-1-4614-8132-4   ISBN 978-1-4614-8133-1 (eBook)
DOI 10.1007/978-1-4614-8133-1
Springer New York Heidelberg Dordrecht London

Library of Congress Control Number: 2013944655

Cover illustration: Courtesy of Lynette Cook, © 2005

Printed on acid-free paper

Springer is part of Springer Science+Business Media (www.springer.com)

For my wonderful and long-suffering wife Nikki, and for my children Aleister, Oran, Arabella, Genevieve, and Vincent

Preface

Initial ideas regarding the habitability of extrasolar planets have focused on the overall size of the planet and its location within the stellar habitable zone. However, many additional factors exist that affect their potential ability to harbor life. This is particularly true of planets orbiting red dwarf stars, for it is these worlds that will have the most time to evolve within a star's Goldilocks zone.

Red dwarf stars live for hundreds of billions or trillions of years, providing a very steady candle with which to light their retinue of worlds. However, red dwarfs throw up a wealth of seemingly contradictory conditions that grossly affect whether a planet will be habitable. Planets orbiting these little crimson suns have a unique set of problems that could, in principle, prevent or reduce their potential to host life. Are the most habitable worlds discovered to date really as they appear?

This book follows the ongoing work of researchers studying the many disparate branches of science, including astrophysics, chemistry, geology, and biology. Their aim is to create a more holistic view of habitability, based around a large number of interlinked factors. In essence, the question of habitability cannot be reduced to an answer based solely on the location of a planet within or outside of its star's Goldilocks zone. This book first explores the nature of the stars and then the planets that orbit them. By considering all aspects of planetary existence, a picture then emerges as to how habitable any planet orbiting a red dwarf can be.

In the making of this book, there were a number of different areas of science that needed investigation. As such, I would like to offer particular thanks to Greg Laughlin (University of California, Santa Cruz), with whom I had many productive discussions regarding the evolution of the lowest-mass stars. Notably it was clear that inexplicably little if any work had been published on the evolution of the orange (K) dwarfs. Aside from one publication in 1993, this area of astrophysics hasn't been visited since. The evolution of these stars was then inferred from work carried out by Greg on red dwarfs, as well as a diverse set of publications on a set of stars called extreme horizontal branch stars. Greg provided some useful clues that validated my assumptions. This is an area of research to which astrophysicists must return. This is all the more true as K-class, orange dwarfs have become targets for searches of habitable worlds.

I would also like to offer thanks to the team at www.universe-today.com for their prescient (verging on spooky) ability to publish articles on topics I was writing about at the time. The work of this website's authors often alerted me to very recent research that was either unpublished or had just been so.

Finally, the book has many areas that are as yet untested hypotheses. In the coming years, Kepler and other telescopic probes will begin to test these ideas, adding meat to the bones of contention and theory. These are exciting times.

Sherwood, UK   David S. Stevenson

About the Author

David Stevenson studied molecular biology at Glasgow University where he attained a First Class BSc. Honors. He then continues is studies towards a PhD in molecular genetics at Cambridge. Further qualifications in with the Open University include a Distinction in Astronomy and Planetary Science, and separately Geophysics and Geochemistry. His peer-reviewed biological research articles from 1999 to 2003 include a paper on the early development of life, "The Origin of Translation," published in the *Journal of Theoretical Biology*.

David's interest in astronomy was encouraged from an early age by his father. This (combined with an interest in explosions!) has led David to research and write about the life and death of stars.

After a stint in academia, David became a teacher but continued to write scientific articles for various publications. He has published numerous articles on the Blackwell Plant Sciences website (2002–2007). "Turning Out the Lights" (an article about red dwarfs) was published in *Popular Astronomy* in 2003, "A Bigger Bang" (about Type Ia supernovae) in *Sky & Telescope* in July 2007 and "Supercharged Supernovae" (cover article in Sky & Telescope, October 2011). "He is currently completing a second book for Springer (Extreme Explosions" which is due for publication in September 2013.

David lives in Nottingham in the UK with his wife and family.

Contents

Part I
Common Themes

1. The Discovery of Extraterrestrial Worlds

Introduction

Spinning wildly, with hundreds of revolutions per second, is the millisecond pulsar PSR B1257 + 12. Not the most glamorous-sounding object in the known universe. Yet any millisecond pulsar has led an extraordinary life. Its first breath is taken in the multibillion-degree plasma generated in the heart of a collapsing star. As the star falls apart, the core is crushed into a fast-spinning ball of neutrons, iron and elementary particles. This is a pulsar, a spinning neutron star. The pulsar generates beams of electromagnetic radiation that sweep outwards from its magnetic poles, scanning surrounding space like the silent beams of a distant lighthouse.

However, at the point of stellar detonation this freshly minted neutron star doesn't cavort with a millisecond pulsar's wild abandon. At this stage, as the debris of the shattered star clears, the pulsar spins slower than a conventional washing machine, perhaps at a few dozen times per second. Over time, normally these pulsars slow and fade from view.

The forerunner of a millisecond pulsar has a different life ahead. Sharing space with a companion star, the intense gravitational field of the neutron star whisks material away from the companion, forming a disc of material around its waist. This accretion disc steadily adds mass to the neutron star. As the pulsar gains mass and momentum, the incoming material spins up the neutron star until it rotates at hundreds of times per second. With increased vigor, the neutron star is reborn as a millisecond pulsar. Its gyrating beams of radiation now rapidly erode what remains of its once vibrant companion.

Pulsar PSR B1257 + 12 wasn't done yet. Rather than sweep the withered remains of its companion under the cosmic carpet,

D.S. Stevenson, *Under a Crimson Sun: Prospects for Life in a Red Dwarf System*, Astronomers' Universe, DOI 10.1007/978-1-4614-8133-1_1, © Springer Science+Business Media New York 2013

the considerable gravity of PSR B1257 + 12 gently nudged what remained into stable orbits. Over the course of the next few million years, the gas-depleted wreckage of the former companion star were molded into a handful of planets. Liberating no detectable radiation of their own, these ghosts of the former Sun betrayed their presence through their subtle gravitational interplay with the pulsar host.

Many millennia later, Aleksander Wolszczan and Dale Frail noticed that the flashes of radiation from this millisecond pulsar varied slightly, as if it was gently but repeatedly being pulled in different directions. Rather than a steady stream of blips, slight variations in arrival time meant that something, or rather some small things, were orbiting it. Judging by the mass these were planets. Not surprisingly, given the nature of the star around which these planets were orbiting, many in the astronomy community were more than a tad skeptical. In part this was down to history. A year earlier Andrew G. Lyne had announced the presence of a planet orbiting PSR 1829–10. However, this paper was later retracted, leaving an atmosphere that was perhaps not surprisingly suspicious. Yet in the case of Wolszczan's and Frail's discovery, the data was sound: planets did orbit this defunct star. Wolszczan and Frail had bagged three firsts: the first confirmed extrasolar planets, the first multi-planet system, and finally the first super-terrans – planets with only marginally more mass than Earth.

Pulsars, odd planetary hosts aside, hold one further record: the oldest planetary system known. Orbiting within the dense stellar core of the globular cluster M4 is a pair of dead stars, PSR B1620-26 and its companion white dwarf. Announced in 2003, this 12.7 billion-year-old system holds a planet with twice the mass of Jupiter in a distant orbit around both corpses. Although this ancient world was probably born in an orbit around the progenitor of the white dwarf, PSR B1260-26b is now locked in orbit around both stars. A distant, peculiar world.

In 1989, a few years before a skeptical audience accepted the presence of Wolszczan's and Frail's pulsar planets, three Canadian astronomers had delivered what would become proof of principle – a successful method used subsequently to detect hundreds of other worlds. Bruce Campbell, G. A. H. Walker and Stephenson Yang used a technique known as radial velocity (described below) to look for the gravitational effect of a planet as it orbited its star. The technique involves looking for the subtle to-ing and fro-ing of the spectrum of a star as a planet moves it gently towards then away from an observer with each orbit.

The presence of an orbiting Jupiter-sized world was implied by what was then cutting-edge spectroscopy. However, the signal was just on the edge of what could reasonably be detected above instrumental noise, and to add to their woes the astronomers had thought that the star was a giant, giving them a misleading impression of its mass and hence companion's planet mass. Gamma Cephei Ab was one half of a binary system. Most skeptics assumed the team had measured not the effect of an orbiting planet but rather the orbital period of both stars in the relatively poorly classified binary. Therefore, for over a decade, the presence of this world lay in limbo. It wasn't until 2003 that improvements in the radial velocity technique finally confirmed that a Jupiter-mass planet orbited Gamma Cephei Ab with a period of 2½ years.

Throughout the early 1990s refinements in spectroscopy and computer technology led to the discovery of the first planets around more conventional stellar partners. Much like water escaping through the crack in a dike, the first discoveries were faltering, hesitant affairs with a relatively long hiatus between each announcement. But by 2000, discoveries were monthly, then weekly, until a total of over 400 worlds were banked. Initially, the majority of these came through the radial velocity technique used by Campbell and described below. However, as technology has swept forward with increasing pace, more difficult techniques have come to the fore. In this first chapter, we examine the techniques and many of the milestones of exoplanet discovery over the last 20 years.

Radial Velocity: The Pull of Extrasolar Planets

In most stellar spectra, chemical elements betray their presence with fine absorption features. As a star and any planets orbit their center of gravity, the pull of any orbiting planet causes the star to wobble in space. This is associated with minute accelerations

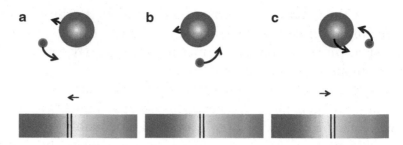

FIG. 1.1 The radial velocity (or RV) method of planet detection. Utterly simple in its science yet fiendishly difficult to use, the technique relies on the precise and accurate interpretation of stellar spectra. Both the planet and star swirl around their common center of gravity. The hidden planet reveals itself by the subtle motion of spectral absorption features to the *red* or *blue* end of the spectrum. Movement to the *blue* end occurs as the star moves towards the observer, and to the *red* end as it moves away

towards and away from the observer as the star orbits its center of gravity. In turn, this effect is manifested as backwards and forwards migrations of stellar absorption lines relative to their resting spectral location (Fig. 1.1).

The amount of wobble varies with two parameters: the relative masses of the star and the planet and the distance between each. The underlying physics is simple. This changes the radial velocity of the star and gives its name to the technique. Although the technique is simple in principle, it does require prolonged and precise observation of the absorption features in the stellar spectra. Where a planet is massive and orbits its star tightly, the wobble stands out from the background noise (Fig. 1.2).

The problem with the technique is that the stellar wobble is minute, perhaps 12 m per orbit of a Jupiter mass world in a tight orbit around a Sun-like star. If you want to find a smaller world in a more distant orbit then your resolution – the ability to discern variability above background noise – must be correspondingly greater. An Earth-mass world pulls with less force than is required to shift its star by a meter. If the orbit of this planet lies within the habitable zone of its star, this weak wobble is extended over 12 months. As radial velocity measurements depend on the precise determination of the location of stellar absorption lines, any interference from atmospheric turbulence or instrument error

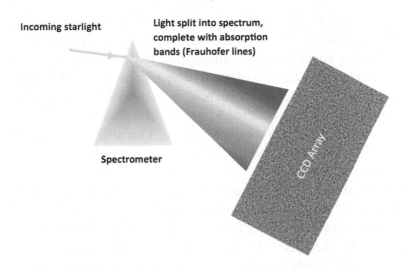

Incoming starlight

Light split into spectrum, complete with absorption bands (Frauhofer lines)

Spectrometer

CCD Array

FIG. 1.2 A graphic representation of how the radial velocity system operates. Incoming starlight is split into a fine spectrum. This is compared to a reference spectrum on a separate CCD array. As the star swings around the system's center of gravity the shifts in the position of absorption (Fraunhofer) lines are detected on the array. This subtle shift (less than one pixel in width) is sensed, and the mass of the planet then determined through the amount of shift. These measurements are repeated for several planetary orbits around the star. In order to operate with high precision the system must be extremely stable and free from any random motion that generates noise

can result in spurious identification of planetary signals. As all of the current instruments deployed to use this technique are Earth-bound our atmosphere presents the biggest problem in the successful isolation of planetary signals.

Furthermore, although a single world orbiting a star produces a nice, clear wobble, additional worlds lying further out produce more subtle resonances that become harder to detect as the number of worlds increases. The number of measurements is of necessity increased, and the sensitivity of these measurements must be greater. Unfortunately, this can lead to noise swamping the signal – or spurious signals emerging from the noise. A common method of avoiding this problem is the use of a reference spectrum. Oxygen was proposed some years ago, but it is the use of an iodine reference that was most widely employed until the rise of HARPS (discussed below).

In the iodine reference model, astronomers pass light through a vessel containing iodine, and the resulting spectrum is compared to the stellar spectrum. The absorption features of iodine are well characterized and can be compared with absorption spectra from the star to root out instrumental noise. Much of the detected variability is caused by atmospheric interference or by thermal motion within the instrument itself. The use of the reference source allows some of the instrumental noise to be tracked and its influence reduced.

In the mid-1990s the radial velocity technique was sufficiently well tested, and its sensitivity improved so that it could be deployed effectively in the field. In 1995 Michel Mayor and Didier Queloz published the first confirmed extrasolar planet orbiting a conventional Sun-like star, 51 Peg b. The word conventional was somewhat loosely applied and only to the star. This was a planet approximating Jupiter in mass, but orbiting its star so tightly that it broiled at over 700 °C. 51 Peg b was the first "hot Jupiter."

Over the ensuing years, subsequent discoveries revealed that these hot Jupiters were really rather common. Perhaps it was our *cold* Jupiter that was the exception. The radial velocity technique is by definition very selective. Given often limited observation times, and the (initially) restricted resolution, the radial velocity technique was always bound to identify the freakish, tightly orbiting worlds. After all it is these worlds that have the most immediate impact on stellar motion and hence spectra. The (initially) low resolution of this technique just couldn't identify smaller worlds as readily as it could the larger, hot-Jupiter mass planets.

Over time the precision of the instruments used to detect planets has improved. In the late 1980s a planet had to swing by its star by more than 10 m/s to stand a chance of discovery. In 1993 Keck's HIRES echelle spectrometer began operation. Conceived by Steven Vogt, this instrument was to capture half the subsequent planet finds, using the RV method. To give an idea of the sensitivity of HIRES and competing instruments, the captured spectra will show variation in the location of spectral bands less than 1/1,000th of a pixel in its CCD array. This sensitivity was improved further with an upgrade in 2003 that allowed HIRES to detect the stellar wobbles caused by Neptune-mass worlds orbiting within the radius of Earth around the Sun. HIRES would later go on to capture evidence for three of the four confirmed super-terran planets orbiting Gliese 581. HARPS would complete the job.

The European Space Observatory's 3.6-m telescope at La Silla in Chile is home to HARPS (*H*igh *A*ccuracy *R*adial *V*elocity *P*lanet *S*earcher) and takes the idea of using a reference spectrum to constrain resolution to another level altogether. Starting in 2003, HARPS employed thorium rather than iodine to produce its reference spectrum. However, to reduce noise, the whole system is chilled to 0.01 K with liquid helium – a fraction of a degree above absolute zero. This super-chilled spectroscope then sits snugly within a vacuum chamber. The combination of minimal atmospheric disturbance and limited particle motion at these low temperatures ensures that HARPS can resolve stellar wobbles down to 30 cm/s – enough to resolve Earth-like worlds in sufficiently tight stellar orbits, or super-terrans in more distant ones. Whereas most RV systems are limited by instrument noise, the extreme localization of the HARPS spectrograph means that it is the idiosyncrasies of the stars themselves that restrict HARPS's unique precision, not the instrument.

Directed by Peg 51b's discoverer, Michael Mayor, and accompanied by Didier Queloz and Stéphane Udry, HARPS holds a number of notable finds, including the first potentially habitable planet GLIESE 581d, added to the three siblings found using HIRES at Keck. Currently, it is HARPS data that forms the basis for argument and counter-argument regarding the existence of another habitable world – Gliese 581d's sibling, planet "g." More on this later in the book.

Through the late 1990s the first planets were found around red dwarfs. Gliese 876b was the first of these, a rare Jupiter-mass planet in a tight orbit around its red dwarf host. In 1996 the first multi-planet system orbiting a main sequence star was confirmed using the radial velocity method, Upsilon Andromedae. It contains three planets, all of which are Jupiter-like. Planets b, c and d were announced in 1996, 1999 and 1999 respectively. With the exception of planet d, all orbit within 1 A.U. of their host star. Planet d, with an orbit of 2.54 A.U., would place it in our Asteroid Belt, making it at best lukewarm and therefore one of the first worlds to be found that wasn't roasting in the light of its host.

Another first for the radial velocity method came in 2001 with the discovery of Iota Draconis b – the first exoplanet found orbiting a red giant. Technically an orange (K-class) giant, the survival of Iota Draconis b confirmed that the rise of the red giant phase didn't mean the annihilation of the entire star system. Not surprisingly, planets far enough removed from the expanding giant can hang on. Iota Draconis b is a particularly massive planet in an eccentric orbit that carries it on average 1.3 A.U. from the center of its star. The orbital eccentricity, coupled to its high mass, meant that its orbital signature was easy to separate from the effects of stellar pulsations, something giant stars are prone to.

In the early 2000s the radial velocity technique had been refined to allow it to detect first Neptune-mass worlds, then super-terrans. The first of these was the second of a pair of planets found orbiting the star Mu Arae c. In August 2004, the HARPS team discovered a planet orbiting Mu Arae with a mass of approximately 14 times that of Earth. Assuming a composition similar to the giants in our Solar System it was most likely a hot twin of Neptune. However, a rocky composition would make it a particularly massive super-terran – a rocky planet resembling a scaled up version of Earth.

Two important scientific advances allowed astronomers to detect planets with masses marginally greater than Earth's. Not only had technology improved sufficiently to detect them, but there was also a reappraisal of the habitability of red dwarfs. This meant that planetary hunts refocused on these small stars rather than the heliocentric pursuit of Earth-like planets orbiting Sun-like stars. With Gliese 876b already in the bag and a hint that other, smaller worlds were also present in the same system, the hunt was on.

Red dwarfs have one big advantage over more massive stars. The center of gravity between the star and its orbiting world is located further from the center of the star than it is with a more massive, Sun-like star. Thus a smaller planet has correspondingly bigger pulling-power around a red dwarf than it would have if it orbited the Sun, and the radial velocity method is more able to detect such a planet in orbit around even a low-mass red dwarf.

Although astronomers waited for technology to improve further, potentially habitable worlds could be found around red dwarfs. Most significantly the orbital period of a habitable planet – the time a planet takes to orbit its star – is small if the star has a low mass and low luminosity. Red dwarfs, therefore, make ideal candidates to search using the radial velocity technique. With the increasingly positive appraisal of red dwarfs as candidate-hosts for life-bearing planets, the radial velocity method took on a renewed vigor as it was used to search for terrans and super-terrans – planets with masses approximating Earth, in the stellar habitable zone.

Transit: The Shadow of a Planet Cast by Its Star

The transit method is a search for the shadow of a moth in a car headlight – when the car is a few hundred meters away. Ground-based observations are effectively worthless. Variations in stellar brightness caused by the transit of a planet across the face of its star are readily swamped by the effects of clouds, atmospheric turbulence – or perhaps the occasional moth looking for food in the night.

Yet, take a probe high above the atmosphere of Earth, lock it securely so that it faces steadfastly in one direction, and you've got the perfect instrument to look for transiting worlds. Although the Hubble Space Telescope had successfully identified the planet HD 209458b using the transit method in 2004, this was a piggy-back discovery. The planet had already been found using the radial velocity technique. That said, Hubble not only confirmed a proof-of-principle, but it also showed that sufficient resolution might be obtained from transit data to determine part of the composition of a planet's atmosphere. Some of the light passing through the planet's gases was absorbed, yielding precious additional spectral clues that could be subtracted from the stellar glare.

Lurking within the data was the spectral signature of sodium vapor. HD 209458b orbits its host Sun so tightly that its broiling atmosphere is boiling off into space, leaving a comet-like tail of hydrogen-rich debris enveloping the planet. Later analysis confirmed that the sodium vapor was found within the extended hydrogen-rich atmosphere at a level corresponding to the planet's stratosphere.

A year later, the Spitzer infrared telescope captured the infrared radiation emitted by two planets: the first of these was HD 209458b, and the second, another hot Jupiter, TrES-1. HD 209458b was observed by Jeremy Richardon (Goddard Space Flight Center) over a range of 7.5–13.2 μm (millionths of a meter). In both cases the emission spectrum provides far more detailed information than the limited absorption spectrum captured by Hubble. Detailed spectral emission lines indicated hydrogen and carbon monoxide, but oddly not water vapor, which had been expected. There was an additional strong peak at 7.78 μm, which was unexplained, and a further peak near 10 μm, attributed to silicate dust – vaporized rock. Later in the same year a separate group identified water using a slightly different technique. These studies opened the gateway on the study of exoplanetary atmospheres and prepared a pathway to the investigation of habitable exoplanets.

In 2007 the transit method was used to identify further molecules within the atmosphere of a distant world. NASA's Carl Grillmair used Spitzer to observe HD 189733b. Analysis of the atmosphere of this hot Jupiter revealed the presence of water and methane. Although it was clearly not a planetary abode for life of any sort, the presence of both methane and water meant the basics of the chemistry of life were present. Indeed, this was the first detection of methane in the atmosphere of any planet outside our Solar System and laid down an important foundation for the future detection of hospitable worlds through their atmospheric chemistry.

Important though Hubble and Spitzer's discoveries were, they would always be limited in scope. Both observatories are involved in multiple fields of investigation. What astronomers needed were spacecraft dedicated to planet discovery. This venture would come with the launch of Kepler, CORoT and the initiation of ground-based systems such as WASP in the ensuing years. Of these Kepler has certainly grabbed the largest share of attention, but all three ventures have revolutionized the field of planetary discovery.

Kepler was launched in 2009 after a 2-year delay. The Kepler telescope swung into position 1 million kilometers from Earth, in a heliocentric orbit, where the Sun and Earth's gravitational pulls are balanced. Without the erratic pull of Earth and the Moon the craft's cameras then focused with unique precision on star after star, detecting the subtle variations in starlight caused by transiting worlds.

With a steady gaze, Kepler has turned the science of planetary discovery into the banality of working in a cannery. A little harsh, perhaps, but effectively true. With its unblinking 0.95 m-wide eye, Kepler gazed at a small square of sky, filled with over 150,000 stars in a field 115 square degrees. Every minute of every hour, its sensitive CCD systems captured the subtle variations in stellar brightness caused by the ephemeral moths drawn to their eternal flames. Of course, the moth-candle analogy is a poor one. A moth fluttering in a car headlight, when the car is a kilometer away, might be a better descriptor. The variation in stellar luminosity caused by a planet transiting a Sun-like star is miniscule even when the planet is as large as Jupiter. Scale that along to see the effect of an Earth-sized world and you get the picture. The variation in stellar luminosity caused by an Earth-like world transiting a Sun-like star is as low as 80 parts in a million – barely above background noise. Thus confirmation of such diminutive worlds relies on data from multiple transits. Super-terrans – large terrestrial or water-dominated worlds, or their larger Jupiter-like relatives will clearly impose a more significant variation in the brightness of its star than a smaller world. However, it was initially unclear how commonplace such giant planets were around red dwarfs.

Despite the difficulty of the technique Kepler and a clutch of Earth-based transit observatories began to churn out planet after planet. After ruling out the effects of gravitational microlensing, star spots, flares and other ephemeral phenomena, astronomers were ready to claim a rather large prize: a vast toll of planets.

After less than 1 year of operation, Kepler had spied and confirmed the dimming effect of over 400 planets. Four hundred is a somewhat embarrassing figure – matching the number found by all other techniques in the preceding 15 years. That said, another 400 or so planets likely skulked in the data from other stars, awaiting further confirmation. Marvelous stuff, indeed.

For a long time planetary astronomy was the preserve of theory or of the study of the Sun's worlds. With Kepler, routine discovery has allowed a lot more science to be dissected and refined and the essence of true discovery to be made. Although Kepler grabs much of the media focus concerning planet discovery, WASP

(Wide-Angle Search for Planets) and France's CORoT (from the somewhat forced acronym COnvenction Rotation et Transits planétaires) have turned up a slew of interesting worlds.

WASP (or more accurately super-WASP) is an international collaboration based at the Roque de los Muchachos Observatory at La Palma in the Canaries and the South African Astronomical Observatory, with headquarters in Capetown. Utilizing a series of wide-angle lens cameras, and with observatories in two hemispheres, a whopping 500 square degrees of sky is covered, extending down to 15th magnitude objects. This allows for a very detailed and broad sweep of the heavens for transiting exoplanets.

Amongst, WASP's notable discoveries are a series of super-heated planets in tight orbits around their host Suns. WASP detected the hottest known exoplanet using the transit method. WASP-33b is a giant planet in an eccentric orbit around its F-class main sequence star. A combination of a tight orbit – at less than 7 % that of Mercury-Sun distance – and a relatively hot central star mean its cloud tops sizzle at over 3,000 °C.

More bizarre still is WASP-12b. Initially crowned as the hottest exoplanet, with cloud-top temperatures in excess of 2,200 °C, WASP-12b was shown to orbit its star so closely that tidal forces distort it into an egg shape. The combination of extreme heating and nose-touching proximity to its star means that this egg-shaped world is being steadily torn apart.

Although evaporating exo-Jupiters had already been spotted – the first, HD 209458b, by Hubble – WASP 12b was a different kind of beast. Whereas the massive HD 209458b would probably survive in some form for the duration of its star's main sequence life, WASP-12b was being shredded at such a rate that it probably wouldn't last much more than 10 million years. To observe the final fling of this planet's short remaining life might have been serendipity in the extreme, or perhaps more likely WASP-12b had suffered some form of interaction with another unseen world, which pulled it into an unstable orbit close to its host Sun. Whatever, the true reason, WASP12-b hasn't much time left. Within another 10 million years, all that will remain of the once-giant planet will be its spectral signature in the gases that make up the star's corona. Ashes to ashes.

However, the transit method has another use. It can detect other unseen worlds orbiting the same star. Where more than one planet is present the gravitational tug of the second, third or more planets will alter the period of time between transits of the known world. This method was successfully employed by WASP to infer the presence of planet WASP-3c, after close scrutiny of the transit times of the transiting world, WASP-3b. WASP-3c does not transit its star, but its presence can be known from the observed effect of WASP-3b. Moreover, where a distant planet orbits a pair of closely orbiting binary stars, the same principle can be applied, and variation in the intervals between eclipses can similarly be used to infer the presence of unseen planetary bodies.

Alongside WASP, CoRoT has been diligently carrying out transit sweeps since 2007 and in essence paved much of the way for the later and heavily delayed Kepler mission of NASA. Not only did CoRoT illustrate that the transit method was an efficient means of planet hunting, it also refined the extent of stellar variability in Sun-like and lower mass stars. This important function constrains a considerable fraction of the background noise, which might otherwise interfere with transit searches. The periodic dimming and brightening of a star caused by a transiting world can readily be mimicked by a star spot rotating in and out of view, or the effect of weak stellar pulsations. In this regard CoRoT demonstrated that there was a greater than expected level of pulsation in many Sun-like stars; information that would then limit the ability of Kepler to carry out its task.

By 2011 CoRoT had bagged around 600 planetary candidates, including the first found to show a secondary minimum in the light curve of the parent star (Fig. 1.3). This occurs as the illuminated planet (CoRoT 1b) moves behind the parent star and the additional reflected light visible on Earth is lost behind the parent star. Capturing a secondary minimum is quite a feat, given the minuscule amount of light reflected from the orbiting planet.

In 2010 CoRoT unveiled COROT-9b. This transiting planet was the first known to enjoy a temperate orbit around its host star. The relatively low mass of its orange, K-class host star (Chap. 3) means that the surface temperature of CoRoT 9b lies somewhere between −20 and 160 °C. This is despite CoRoT 9b orbiting at an equivalent distance to Mercury. At its time of discovery in 2010, the orbital separation of 0.36 A.U. was by far the largest of any transiting exoplanet from its star that had been observed.

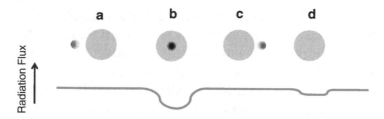

FIG. 1.3 The (exaggerated) effect of a transiting planet on the light curve of its parent star. As the planet moves in front of the star (*B*) the light is partially blocked, causing the star to appear to dim. This is what Kepler is looking for in its field of 150,000 stars. More rarely, a much shallower secondary minimum is detected (*D*) as the illuminated planet moves behind the star and less light is reflected in the direction of the observer. In reality the light curve will be more complex as the area of the planet illuminated by its star varies throughout its orbit. CORoT was the first craft to observe this effect with CORoT 1b

At the other extreme lies HD 14902b. This world was initially identified in 2005 in radial velocity data by the N2K Consortium. The consortium was a 2-year-long collaborative venture of Chilean, Japanese and U. S. astronomers to identify extrasolar planets using the radial velocity technique. The project focused on 2,000 nearby bright main sequence stars that were not already the focus of investigation in the astronomical community. Observations ran at Keck, Subaru and Magellan, plus an automated 'scope at the Fairborn Observatory, which looked for planetary transits.

The expectation was that around 60 planets would be found, but in the end only seven were confirmed. Of these HD 14902b was rather unusual. Each transit decreased the light from its host star by 0.003 magnitudes. This doesn't sound like much, but it was enough to allow the amateur astronomer Ron Bissinger to confirm it. Moreover it suggested that HD 14902b was a fairly large planet, close to its star. Once more the combination of data allowed a lot of detail to be gleamed concerning this odd world. The concomitant identification by an amateur astronomer using a backyard 'scope equipped with its own CCD array opened the door of planet discovery to the amateur astronomy community in general – a huge achievement.

Despite being yet another hot Jupiter, HD 14902b stands out in that it is extremely dense. More than three-quarters of its 114 Earth-mass bulk is rock and metal – a whopping 80–110 Earth-masses.

The presumption is that this odd world is the torched remains of something even bigger. Bombarded by intense radiation from its host star, much of its volatile material has boiled off into space, leaving the rocky core flanked by a streamlined gaseous layer. HD 14902b (and CORoT-7b) is now regarded as a Chthonian world – a term coined from the Greek term for underworld. These blistering planets are often massive, but the loss of volatile materials leaves them grossly enriched in much more refractive, dense material. They would not make comfortable abodes for life.

As resolution has improved the transit method has evolved to provide additional information. In 2011 Kepler spied on a very tight clutch of worlds orbiting the star Kepler 11. This tight nest of relatively low-mass planets has five orbiting within the orbit of Mercury, were they somehow transposed to our Solar System. The sixth lies somewhat further out, nestling in between what would again be the orbit of Mercury and Venus in our system. Planets Kepler 11b and c orbit with a 5:4 orbital resonance, implying a strong gravitational interaction between them. Aside from the very dense orbital arrangement – the densest seen to date – this was the method by which the masses of the planets were determined. All six planets transited Kepler 11, giving precise parameters for orbital separation and angular size of each world. On its own a transit can't be used to determine mass. However, combine several transits from planets in the same system and a little bit of magic can be accomplished.

This transit-timing variation (or TTV) method relies on precise measurement of the transits of the known world or worlds. As each world orbits close in towards the central star, its individual gravitational pull either delays or advances the transit time of neighboring worlds. With a little help from Kepler's Third Law of Motion, the distances to each planet can be determined and the relative mass of each world inferred. A little bit of high school math and the parameters of each of Kepler-11's worlds were in the bag.

The fairly routine mathematics reveals a lot about the Kepler-11 system. The planets all orbit within a few degrees of the same plane – although Planet e appears to deviate a little more profoundly from this. All of the planets appear to be of low density, implying that despite their tight orbits, they were born with a fairly large proportion of volatile (light) elements.

This is significant. These light elements are predicted to be limited in abundance close into the developing star, as the high temperatures would tend to boil them away. By implication, each of the Kepler-11 worlds must have been born further out from the protostar and then migrated inwards, bringing their inventory of volatile elements with them. Not only does this pattern of orbital migration tally with theorized events in our Solar System, but it also means that the planets that form around red dwarfs may still be volatile-rich despite their torrid beginnings. Volatile elements are essential for life; thus the Kepler-11 system constrains some of the parameters needed for the future habitability of exoplanets. More on this later.

The beauty of this technique is that you don't need to observe a transit for each planet. With enough data, an unseen, non-transiting world can have its mass determined through its effect on those worlds that are observed.

The transit technique's advantage over the radial velocity technique lies in its ability to refine planetary diameter and resolve planetary atmospherics. This undoubtedly impacts on our understanding of habitability. However, transits only occur when the orbital plane of the planet is aligned with our line of sight. Clearly this is a limitation, as simple geometrical arguments imply that the majority of stellar orbits will fall outside those that transit the star. Compared to Kepler 11, most planetary systems are more spread out. With longer periods of time separating each transit, the TTV method is more difficult to apply. The gravitational influence of each planet on one another is proportionately less significant, and the length of the transit too drawn out. Thus the impact of any additional planet on transit time is more readily lost in instrument noise.

Moreover, transits are limited by the relative diameter of the star, its planet and the distance between them. Where stellar variability is great, the effect of star spots and flares also has the potential to obscure the dimming effect of transiting planets. With an increasing focus on red dwarfs these limitations are more to the fore. Red dwarfs, more than any other type of star, are prone to the youthful problems of spots and flares. The detection of transiting worlds is consequently more complex. A transit is a useful thing, but it isn't perfect.

Yet, astronomers have found yet another route around the problem of determining planetary characteristics even where the planet's orbit doesn't carry it across the face of its star. In 2012 a novel technique was published in the journal *Nature* that circumvented this limitation. The star system Tau Boötes was known to host a planet. τ Boötis b was one of the first identified using the radial velocity method. However, there was considerable uncertainty as to the nature of the planet's orbit and hence its mass. Moreover, without available transit data, no orbital information or planetary atmospherics could be directly determined for τ Boötis b. To get around this Matteo Brogi used the Cryogenic Infrared Echelle Spectrograph (CRIRES) at the Nasmyth A focus of the Very Large Telescope UT1. This unique infrared spectrograph is located at the European Southern Observatory (ESO) on Cerro Paranal, Chile.

Scanning the system over many days, as the hot Jupiter swept rapidly along its orbit, the CRIRES spectrograph detected the clear and varying signature of carbon monoxide. This noxious gas wasn't present in the atmosphere of the star. It had to be in the atmosphere of τ Boötis b. The thorough baking the dayside of the planet received from its host star meant that the planetary carbon monoxide readily emitted a potent spectral signature.

As the tidally locked planet swung around its star, the carbon monoxide hot spot appeared and disappeared, as first it was blocked by the star and later as the view of it was occluded by the planet itself. Using the periodic signal of the carbon monoxide hot spot, Brogi and co-workers showed that the 5.95 Jupiter-mass planet orbited its star at an angle of 44.5°, carrying it high above the line of sight needed for a transit to be observed from Earth. Not only was this ingenious method able to bypass the block caused by a lack of a direct transit, but it also opened the door on the discovery of planets using direct infrared emission. The data had such high resolution that Brogi and colleagues were able to show that the carbon monoxide was concentrated in the planet's lower atmosphere. The presumption was that this noxious gas was destroyed at higher altitudes by ultraviolet emission from τ Boötis.

Interestingly, these observations contrast with those of other hot Jupiters where ther appears to be a lid on atmospheric motion. In tau Bootis b temperature inversion limits atmospheric motion and therefore restricts the movement of carbon monoxide from deeper within the planet to the upper atmosphere. In other hot Jupiters, carbon monoxide concentration peaks at greater altitudes. This indicates that in the atmospheres of these worlds gas motion is more restricted. It also suggests that the atmosphere is less heavily irradiated with ultraviolet light from their parent star. The latter effect often varies inversely with stellar age. Older stars have fewer star spots, and this, in turn, means that there are fewer accompanying stellar flares. It is the flares that generate the preponderance of ultraviolet light in Sun-like stars.

Kepler does have one limitation. Its eye scans its patch of sky in the northern hemisphere. While the Kepler instrument looks north for stellar twinkles, the impressive HARPS is busy looking the other way. Where radial velocity and transit data are combined a lot of useful extrasolar planet science can be done on. The radial velocity method constrains mass directly, as well as orbital parameters. The transit method defines planetary diameter and can provide information on the chemistry and dynamics of planetary atmospheres. Having parallel systems would clearly improve our understanding of alien worlds. As such, at the time of writing there are moves to build a HARPS-North instrument that could then collate data with the Kepler mission, which identifies planets through transits (discussed later). Plans for this parallel system are far from complete, and it will be a number of years before HARPS-North is functional.

Meanwhile, Kepler will keep scanning its patch of planetary heaven until at least 2015, barring any system failures.[1] Even without HARPS working alongside it, astronomers will be able to infer the mass of some of their worlds through the gravitational interplay between each as they orbit the common center of gravity and modify the transit times.

Microlensing: The Ghosts of Hidden Worlds

Einstein's general theory of relativity states that matter bends space and the curvature of space alters the path of any incumbent light. Therefore, when a massive object passes in front of a star,

---

[1] While going to press Kepler has broken down – probably for good following the failure of a critical gyroscope needed for the alignment of the craft.

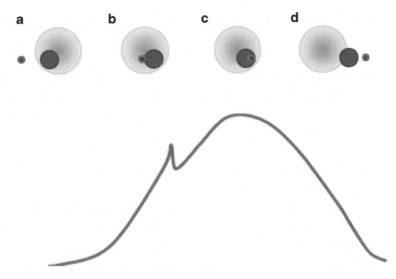

FIG. 1.4 Microlensing of a distant star by a foreground star and its orbiting planet. As the foreground red dwarf moves in front of the distant star, its gravitational field focuses the light from it, causing the light of the distant star to brighten in a predictable fashion, which relates to the mass of the red dwarf. If a planet orbits the star, the additional planetary mass causes further brief magnification (B) on top of that caused by the red dwarf alone. This boost is transitory in duration, as the time taken for the planet to orbit the host star is considerably shorter than the overall transit time. The distances between both stars are measured in hundreds or thousands of light years

the gravitational field of the object focuses the star's light. Much like a magnifying glass, the lensing object causes the star to brighten and then dim again in a predictable way, dictated by the curvature of space. The extent of the change in brightness tells you the mass of the lens and its distance from the star (Fig. 1.4).

During the 1980s there was debate among cosmologists as to the nature of the mysterious dark matter that dominated space yet couldn't be seen. One group suggested that dark matter was comprised of fundamental particles created in the Big Bang. These were WIMPs, or Weakly Interacting Massive Particles. Another camp insisted that dark matter was nothing more than dim objects made of conventional matter, such as brown or red dwarfs, neutron stars and black holes. These were MACHOs, or MAssive Compact Halo Objects.

During the early 1990s, Polish astronomers Andrzej Udalski, Marcin Kubiak and Michał Szymański, and the late Bohdan Paczyński, established OGLE (the Optical Gravitational Lensing Experiment), based at the University of Warsaw. Most of the observations have been taken at the Las Campanas Observatory in Chile. OGLE had two principle successes: the first in November 2002 was to detect the first planet through the transit method, OGLE OGLE-TR56b. Until the detection of WASP-12b in 2008, this hot Jupiter held the record for the shortest orbital period.

Its second "first" came in 2004. OGLE and New Zealand's Microlensing Observations in Astrophysics (MOA) identified a planet orbiting a distant star, taking the prize as the first planet detected using the microlensing method. OGLE-2003-BLG-235L b is a1.5 Jupiter-mass world was found indirectly. Instead a background G-class star was lensed by a foreground dim M-class dwarf (see Fig. 1.2 and Chap. 3). Nothing very exciting there, given the abundance of red dwarfs. However, if the background star was lensed by an unaccompanied M-class star, the pattern of brightening and dimming would have been monotonic – a simple rise and fall lasting a few days to weeks. That was not the case. Instead the light flickered in a more complex pattern that betrayed the presence of a planet-mass object in orbit around the lensing red dwarf. The additional mass led to a secondary peak in brightness lasting a few hours. In subsequent years further planets have been found this way.

In 2006 OGLE announced it had bagged the most distant exoplanet – a record that still stands. The 5.5 Earth-mass planet, glamorously named OGLE-2005-BLG-390Lb, orbits its red dwarf host at a distance of 2.6 A.U. The whole system lies a whopping 21,500 light years away towards the center of the Milky Way. The wide separation of the super-terran planet from its dim red dwarf star also implies that it is the coldest exoplanet currently known.

On its own this technique is akin to looking for a needle in a haystack. However, if you can crank up the numbers and observe thousands of stars frequently, you will identify these rare lensing events, caused by a passing star and its planet. However, once more Kepler can accomplish the same feat – while it's watching for transits. Inevitably, while Kepler stares at its small star-rich patch of sky it will observe lensing events. Although these effects may take a bit more time to disentangle from others, Kepler is ideally suited to identify planets this way, as the number of stars in its field of vision is great.

While Kepler has been looking for dips, the international collaboration PLANET (Probing Lensing Anomalies NETwork) has been scrutinizing thousands of stars across the hub of the galaxy and examining lensing of these by foreground stars. PLANET is a unique network of collaborating, small (1–2 m) optical telescopes operating from bases in Australia, Chile, Denmark, South Africa, the United Kingdom and the United States. Again, the use of sensitive CCD arrays linked to each 'scope allows for the precise detection of minute variations in apparent stellar output caused by microlensing.

In all, PLANET detected only 40 microlensing events, and three of these were previously confirmed finds by Kepler. So, what was all the fuss? Well, simply put, extrapolating the Kepler yield across the number of microlensing events – and the surveyed area of the galaxy – meant that the galaxy must buzz with 100 billion planets, 1,500 within 50 light years of Earth. Digging deeper, the pattern of microlensing events indicated that roughly 15 % of galactic stars hosts a Jupiter-mass planet, 50 % have a Neptune (14 Earth)-mass exoplanet and 66 % have an Earth-mass world. Clearly, the most habitable worlds – at least for complex life – are the most abundant: good news. The logic behind the link between planetary mass and habitability for complex life is explored in depth in Chaps. 9 and 10.

Early analysis of massive planet prevalence clearly showed that they were more abundant around more massive and more metal-rich stars. Clearly there are exceptions – the case of M4's pulsar orbiting giant is one. However, the trend in the data is clear. Massive planets form where gas can cool more readily and there is more of it. Further recent analysis by Lars Buchhave constrained the planetary mass distribution further, using the abundant data from Kepler. In general, high-metallicity stars – those with more abundant concentrations of elements heavier than hydrogen and helium – have more of the massive Jupiter-like worlds, while lower metallicity stars, like Gliese 581 and 667C, are deficient in these worlds. Neptune-mass and super-terran planets show no overall preference for stellar metallicity but are somewhat more common around stars with metallicities similar to the Sun. Therefore, we can't say that there is a further link between metallicity of the parent star and the overall mass of the planets orbiting those stars. However, are low metallicity stars dominated by planets with a greater proportion of lighter elements than those formed from more metal-rich material? This is an important point to consider as the top two habitable exoplanets – at least from the perspective of the stellar habitable zone – are both relatively poor in metals, compared with the Sun.

Now You See It, Now You Don't: Formalhaut B and Beyond

How wonderful it would be to see an extraterrestrial world. Before Hubble raised its eyelid above Earth's atmosphere, and long before the Cold-War opened an avenue of exploration using adaptive optics, all astronomers could do was daydream that one day they might catch a glimpse of an alien world, orbiting serenely around some distant star. Earth's turbulent atmosphere distorted the signal of any such world into a murky haze of scattered inferences. Catching a glimpse of an alien world was impossible.

Then along came first Hubble, an eye poised above the cloud-tops. This was followed by superlative optical systems able to compensate for the minutia of atmospheric motion.

It had been unknown for some considerable time that several nearby A-class (white) stars had discs of dusty material orbiting them. The warped edge-on shape of the debris disc orbiting Beta Pictoris was such that it required the presence of at least one Jupiter-mass planet to mold it. However, until the middle of the last decade nobody had thought to look for a planet. The resolution needed was beyond that thought possible. However, Formalhaut provided a new opportunity. Its disc was nearly face on and was there to be anything sizable hidden within it, perturbations in the motion of material might reveal the presence of the massive, orbiting body. Indeed, after 8 years of scrutiny by the Hubble Space Telescope Thayne Currie (formerly of the Goddard

Spaceflight Center) imaged one "blob" moving 115 A.U. from the central star in what appeared to be a predictable orbit. The blob, apparent in visible light, seemed to follow an orbital path, albeit one cutting through the debris disc. The pattern of infrared radiation detected by Hubble suggested a temperature of a few hundred Kelvin in line with expectations for a giant planet a few tens of millions of years old.

Three years prior to the 2008 discovery, Paul Kalas and James Graham had inferred the presence of such a planet from the pattern of motion of a more distant dusty ring. However, its direct observation was ground-breaking – if true.

Exciting though the discovery was, in 2010 the infrared Spitzer telescope failed to identify the planet at near-infrared wavelengths. Was the planet merely a specter; a mirage caused by a distant stellar flame, scattered through the dusty disc? The Atacama Large Millimeter Array (ALMA) has resolved the edges of rings within the Formalhaut dust disc. The location of these tight edges supports, in principle, the existence of Formalhaut b and a more distant planet. That said the masses implied for each world differ significantly from those implied by the visible-light images. Hubble's images taken in visible light imply a mass similar to Jupiter for Formalhaut b. ALMA suggests the mass is somewhere between that of Mars and a few times that of Earth. Clearly, something is amiss here that will require further rounds of infrared and visible observation.

Finally, in 2012, Currie announced the results of further observations using Hubble. The said planet does indeed follow a path aligned with the debris disc. The mass of the world appears to be less than that of Jupiter, but this is not yet constrained. Unfortunately, as the planet appears to be visible as a result of accretion of surrounding gas and dust, it is not directly visible after all. Consequently it has lost its title as the first directly observable planet. You win some, you lose some.

Nearly simultaneous with the initial discovery of Formalhaut b came the announcement of four giant planets orbiting the more Sun-like star HR 8799. These planets are thoroughly secure. HR 8799 is a 30-million-year-old F-class star with 1.5 times the Sun's mass and just under five times its luminosity. Direct imaging of each planet was accomplished by blocking the light from the central star coronagraph.

The discovery of the first three of the four giant planets was announced by Christian Marois and his team (University of Montreal) in November 2008. Images were taken using the Keck II and Gemini telescopes in Hawaii. The outermost planet swung around its star within a very dense, dusty disc of material. Between this outer disc and a less substantial inner one, lay the other two worlds. The fourth planet was observed at higher resolution a year later.

The system resembles a scaled up version of the outer Solar System. Planets e, d, c and b lie at two to three times the orbital radii of Jupiter, Saturn, Uranus and Neptune. With a more luminous central star, each receives approximately the same radiation flux as the equivalent worlds in our planetary system. However, here the similarity ends. All four planets are veritable giants. Each would put Jupiter in its place were they to move into orbits around Sol. Each world has a mass exceeding that of Jupiter and approaching the deuterium-fusion limit of 13 Jupiter-masses. At this point these planets might be reclassified as brown dwarfs (see Chap. 3) – failed stars. It would seem scaling up the mass of the star scales up the mass of the planets around it. The reasons for this relationship are discussed more fully in Chap. 2 and are returned to in later chapters. In essence a bigger disc of material is needed to build a bigger star. But a more substantial disc of material will also feed the growth of larger planets. This is an important relationship once we get down to the lowest mass stars, red dwarfs.

Nudging both of these optical 'scope discoveries into second and third place was the identification by Daniel Lafeniére of 1RXS J160929.1–210524 using the infrared eye of Gemini in Hawaii. Although it was announced in September of the same year – 2 months ahead of the Formalhaut b discovery – it took nearly two further years of observation before the distant, 8 Jupiter-mass planet was confirmed, orbiting its star at a distance of 330 A.U.

A rather more unusual planetary find was the discovery of a 3.3 Jupiter-mass world, orbiting a 25 Jupiter-mass brown dwarf in 2004. With a temperature of approximately 1,000 °C the planet is visible only in the infrared, but probably glows a dull red through clouds of dust and organic molecules. The two objects orbit their common center of gravity at a separation of 41 A.U. or 41 times the Sun-Earth distance. Both objects appear to be young and still accreting material from a circumbinary disc of material.

One further oddity emerged in the early part of the millennium – the discovery of free-floating planets using infrared observations. These objects have masses approaching that of Jupiter. S Ori 70 was the smallest. Based on its temperature, and presumed age of 5–7 million years, this object appeared to have only five times the mass of Jupiter and was free-floating in the Orion Nebula's star-forming region. Whether this was a true planet, kicked loose from its parent star; a prematurely terminated brown dwarf; or a true planet formed entirely in situ remains unclear.

What Worlds Await Us?

We shouldn't be surprised at the peculiar worlds thrown up by Mother Nature. After all, our first confirmed planetary system consisted of four battered worlds swinging precariously around the post-supernova corpse PSR 1257 + 12. Subsequent finds have proved no less intriguing. Quite aside from the plethora of hot Jupiters, we have been confronted by a Conga line of extravagant planets, some hell-bent on self-destruction, others smothered in deep oceans. Above and beyond these finds are other possible worlds made up of exotic chemicals, found only rarely on Earth. We conclude this chapter with a tour of what is known, what is likely and what is possible.

Chthonian Worlds

As mentioned earlier, Chthonian is the Greek term for the Underworld. We imagine that the Underworld is like a Biblical hell, tormented by flame, an oppressive, dark domain. The Greek term has found itself applied to an odd collection of worlds, each worthy of its place in Hades. These Chthonian worlds take a viciously tight orbit around their host stars and pay the price for their close relationship.

With orbits that take less than a day to complete, Chthonian planets may travel at more than 5 million kilometers per day. Tidally locked to their host star, the permanent day side may broil at over 2,000 °C. At these temperatures rocks are liquefied, and many of their chemical constituents have become gaseous. Many silicates, the building blocks of planets, are found in the gaseous state, giving rise to an atmosphere of boiled rock, rather than the familiar oxygen, nitrogen and carbon dioxide.

CORoT-7b is the current record holder in terms of temperature, with a surface cooked to over 3,000 °C. Unless the mass of the planet is high, any volatile element that we might be used to has long since been boiled off into space, leaving vaporized rock and metal. Such worlds are often relatively massive and show ample evidence of having arrived in hell sometime after birth, rather than being born in situ in such a precarious orbit. HD 14902b is a case in point. With a proportionately massive metal and rock-rich core this world must have been born with a substantially bulkier envelope of hydrogen and other light elements than it currently retains. Two routes could explain its current state. The first catastrophic impacts removed its volatile materials, leaving a stripped-down core. However, with such a massive core, it would seem highly unlikely a collision (or collisions) could be sufficiently violent to remove such an evidently large mass of volatile material. This scenario works fine for Mercury in our system, but Mercury is less than 1/200th as massive as HD14902b.

Instead a scenario involving orbital migration seems most likely. HD 14902b was born, perhaps as far as 4–5 A.U. from its host star. Embedded within a massive disc of gas and dust, frictional and gravitational forces then dragged it inwards towards its star. For whatever reason, be it the loss of disc material, tidal interactions with the parent star, or interactions with neighboring giant planets, HD14902b ended its migration less than 0.1 A.U. from its star. Bombarded by intense radiation and stellar winds, HD 14902b found that it was unable to hold onto its lower density hydrogen-rich envelope. Over the ensuing few million years, most of this was removed, leaving the more recalcitrant core behind, bathed in a stripped-down retinue of volatile material. Other worlds are less lucky.

WASP-12b is on its final legs. With an orbit so tight that tidal forces compound the heating by its stellar partner, the planet is steadily being torn asunder. In less than 10 million years this Chthonian world will evaporate, leaving a brief disc of debris around its star. Soon even this will disperse in the intense stellar gale (Fig. 1.5a).

HD 209458b is a transitional world. The first planet to have its atmosphere sampled through spectroscopy and the first to

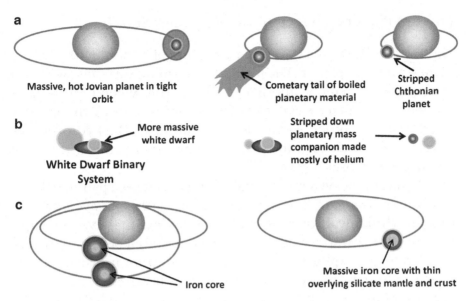

FIG. 1.5 Formation scenarios for three exotic worlds: (a) Chthonian planet; (b) a helium planet; and (c) a "cannonball." In (a), a massive hot Jupiter is irradiated by its star so strongly that the hydrogen- and helium-rich outer layers are boiled off, leaving the massive, rocky core. WASP-7b may be an example. Planets HD 209458b and HD 14902b would represent intermediaries in this process. In (b), two white dwarf stars orbit a common center of gravity, and the smaller but more massive one strips material from its helium-rich companion. This eventually leaves a helium-rich and planet-size mass companion in a fairly tight orbit around the white dwarf star. In (c), two differentiated planets collide early on in their history. The iron-rich cores merge, and much of the overlying rocky mantle is blasted off into space. The resulting planet is much like Mercury, but could, in principle, be much more massive. Stars and planets not drawn to scale

be uncovered through transit – albeit as a piggyback affair – HD 209458b is a torrid world cloaked in a cometary stream of its own material. Although orbiting further from its Sun-like star than WASP-12b, it still receives sufficient radiation to drive off a massive stream of hydrogen and helium. The planet sweeps its orbit much like a comet would around the Sun. Given the rate of mass-loss HD 209458b will survive to the close of its star's main sequence life – albeit in a much more emaciated form. However, like all hot-Jupiter worlds, within a few million years of its star's ascent onto the red giant branch, tidal forces between the stellar envelope and the planet will draw what shattered planetary remains exist into the fiery depths of the star.

Helium Planets: Gravestones for Former Stars?

Figure 1.5b illustrates the formation route of a thoroughly odd gas giant – a helium planet. The formation scenario involves two aging stars. Both orbited one another so tightly that when both became red giants the denser star ended up orbiting within the swollen envelope of the other. As the denser star did so, the hydrogen-rich envelope was cast off into space, and the two helium-rich cores spiraled ever closer together.

In these star systems the end product is either a pair of helium-rich white dwarfs, or if one star was sufficiently massive a carbon-oxygen white dwarf orbited closely by its less massive but larger helium white dwarf partner. These systems often betray their existence through explosive interactions. The more massive carbon-oxygen white dwarf steals helium from its less massive partner as the two orbit one another every few minutes. The helium builds up on the denser white dwarf until it is hot enough to ignite and explode. The resulting explosion is called a helium-nova. These AM Canum Venaticorum (AM CVn) systems are rare, and their ultimate fate is unclear.

In some recent scenarios the final addition of helium from the shriveled companion to the more massive white dwarf triggers a so-called Type Ia supernova. However, at least a few white dwarf binary systems are known that appear to have avoided this fate. If the mass of the helium companion falls low enough the interior of the remains readjusts itself, leaving a small core of highly compressed heavy elements from the original star, an outer core of solid helium surrounded by a region of liquid helium. On top of this will be a mixture of helium gas and any hydrogen that somehow survived, along with a small amount of heavier elements that were present in the original star. If the mass of the "planet" is less than 13 Jupiter-masses or so all the helium will be present in a liquid or gaseous form.

Carbon Planets

Imagine a world made primarily of carbon and carbon compounds. What an odd world this might be. But is such a planet possible, and if so, how might it be formed?

The disc of material that made the Sun and Earth was rich in oxygen – the byproduct of explosive nucleosynthesis – the formation of elements in supernovae. However, there are some stars that produce a much greater abundance of carbon compared to oxygen. These carbon stars are relatively rare red giants on their last legs.

In a carbon star, helium fusion is firing up in fits and starts, and the product carbon is being dredged to the surface of the star by convection – the stirring motion seen in a pan of boiling water. Whereas, most red giants convert most of their carbon to oxygen, carbon stars slough off much of their ashes into interstellar space before this trick happens. Consequently any planetary system that forms from the debris of this material might just be sufficiently enriched in carbon to form some uniquely peculiar worlds.

On Earth, elemental carbon is found in three forms: graphite – the stuff of pencils; Buckminster fullerenes, found in soot; and most glamorously of all, diamond. Graphite and fullerenes are stable at modest pressures. However, once you plough down a couple of 100 km into Earth's mantle graphite magically rearranges its molecular structure to form diamond. Imagine then a planet formed from carbon-rich material. What would its interior be like? Figure 1.6 illustrates the possible internal structure of such a world.

The core would resemble that of our world. Iron and nickel would melt during planetary formation and flow towards the center under gravity as they would be by far the densest materials. Meanwhile silicon, which is bound to oxygen forming silicates on Earth would instead form silicon carbide with carbon. Embedded among this rigid silicon carbide layer, or forming a stratum on top, would be a thick stratum of solid diamond. The diamond layer would extend to a few hundred kilometers below the surface. Above this a graphite-rich layer would form a thick crust.

On Earth hydrogen is bound to oxygen-forming water. A carbon planet would combine its hydrogen with the copious amounts of carbon-forming hydrocarbons – oil. Much like a scaled-up version of Titan, this oily ocean might cover the entire planet, or form isolated lakes or seas, flanked by flat graphite landmasses. Interestingly, analysis of the extrasolar world 55 Cancri e by Nikku Madhusudhan (Yale) suggests that it may be a carbon-planet – or at least one very much enriched in this element, compared to the Sun

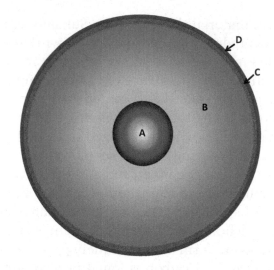

FIG. 1.6 Anatomy of a carbon planet. If oxygen is limiting in a planet's building materials, carbon chemistry may take over. The core, *A*, may be made from the carbon-iron alloy familiar to us as steel. Above this, *B*, may be a silicon-carbide mantle, enriched with abundant carbon inclusions – diamonds. Lying further out at *C* may be a thin graphite layer forming a crust, submerged in a sea of oil (hydrocarbon), *D*

and its worlds. The parent star has a greater proportion of carbon than the Sun, and its planets may thus have a carbon-rich interior. Moreover, the density of the planet seems to confirm this. If so, 55 Cancri e may be a real gem of a planet.

Since both diamond and silicon carbide are extremely tough, the interior, regardless of its internal heat, would be frozen in place. Thus there would be little or no tectonic activity. However, the release of methane, hydrogen or other gases from the deep interior might provide for some violent explosive activity that might just shower the landscape with diamonds, or at least a cascade of crude oil. However, once the interior had degassed, the planet would fall into eternal sleep, its heat slowly escaping by conduction through its diamond depths.

Cannonballs

Imagine a world that is made mostly or exclusively of iron. In our Solar System only Mercury really comes close to qualifying. This dense little world packs a surprising punch. Not only has it clung onto a magnetic field over 4.5 billion years, but data from the Messenger spacecraft clearly indicates that the core is enormous in proportion to its overall size, occupying 75 % of the planetary volume. The mantle is confined to a thin veneer of partially molten rock overlying this layered ball of iron and nickel.

How did Mercury end up this way? There are two plausible and not exclusive alternatives. In the first Mercury was simply born loaded to the hilt with metal. Its location in the Solar System, close to the Sun, would have ensured that the material with low melting points Mercury condensed from was dissipated. This would leave metals, of which iron is by far the most abundant, along with some refractory (high melting point) metal oxides and silicates. Much of the material that forms Earth would simply have boiled away. Modeling of the early Solar System also implies Mercury probably coalesced near the inner edge of a disc of material, so was intrinsically low by mass and depleted in volatile materials.

However, there is a much more violent scenario. Computer modeling of the early Solar System shows an abundance of rocky planetesimals – perhaps 100 or so objects ranging in mass from that of the Moon to Mars. These careered around the inner solar system, augmented by a barrage of lesser bodies, flung this way and that by the coalescing outer planets (Chap. 10). Within this soup of material, collisions would have been commonplace. Earth plays testimony to this. A giant Moon orbits our world closely – the byproduct of a collision between the proto-Earth and a Mars-sized planetesimal, christened Thea.

At Earth's orbital position material from the collision could collect into a disc around the young Earth, ultimately raining most of the debris back onto our world. Close into the Sun, planetesimals move more rapidly in response to the greater gravitational pull, and consequently collisions are more violent. In such an environment a collision between two Mars-sized worlds would drive much of the less dense material into an orbit around the Sun, rather than around the proto-Mercury. The densest material might tend to collect in what remained of the proto-planet, leaving a world enriched with iron but strongly depleted in the lighter rocky material that would otherwise form its crust and mantle.

Figure 1.5 illustrates the formation scenario just described. A "cannonball," or iron-world, forms in response to a violent collision between two iron-rich planetesimals early in their existence. The resulting world may vary considerably in size, depending on the composition of the two planetesimals, the velocity of their impact and the mass and distance between them and their Sun. Cannonballs may well be commonplace, as the means of their creation is not hard to come by. Some cannonballs may in fact be Chthonian planets, stripped down through the impact of stellar radiation to their cores. Even Mercury shows some scars from its continuous battle with the solar wind – craters oddly depleted in fine dusty debris. Imagine, then, a planet orbiting its Sun-like star at less than 1/10th the Sun-Mercury distance. Radiation would be 100 times as intense and the effect on the planet correspondingly greater.

Cannonballs should be detectable through the Radial Velocity and Transit methods. Astronomers will look for any massive planetary body whose size indicates a high density. Expect the announcement of their discovery soon.

Aquaplanets with Ice Mantles

Ganymede is Jupiter's largest and an utterly extraordinary moon. Bigger than Mercury, this frozen world holds an ocean 100 km deep beneath its icy mantle and a fluid iron core deeper still. Transport Ganymede to the orbit of Mars, and its deep mantle of ice would start to melt. The product, albeit a very transient one, would be an aquaplanet 5,262 km across. why transient? Unfortunately the low gravity would allow most of the water to escape into space (Fig. 1.7).

However, what if we made a planet like Earth or larger and bulked up the mass of water to create an ocean 100 km deep. At its deepest point, such a watery-terran would be crushed under 10,000 times the atmospheric pressure at Earth's surface. Compare that to Earth, where the deepest recesses of our oceans are less than 11 km deep. Below a few hundred meters the oceans are dark, cold places, where sunlight is barred from penetrating. Imagine then increasing the depths of our oceans by tenfold. Are there planets like this in the cosmos, and what would conditions in their dingy ocean depths be like?

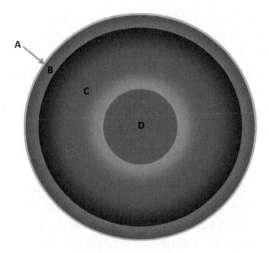

FIG. 1.7 Structure of an aquaplanet. *A* Represents the water-rich atmosphere. Underlying this is an ocean tens to hundreds of kilometers deep (depth exaggerated for clarity here). In *B*, water transforms into a variety of exotic ices in depths exceeding 90 km, depending on the gravitational pull of the planet. *C* Represents the rocky mantle, also rich in water and potentially active, powering undersea plate motion and hydrothermal vents. *D* Represents the metallic core. Ganymede is our closest analog but is obviously much smaller. With increasing mass, these planets grade into ice giants such as Uranus and Neptune

The likeliest answer to the first question is a clear yes. It is highly likely that there are planets with a few times the mass of Earth that are born with enough water to fill oceans to great depths. How great depends on how much water they retain at birth, or acquire later through cometary impacts. There will be more discussion of these processes and their impact on the prospects for life in Chaps. 2 and 10.

Imagine taking a very robust submersible and traversing the great depth of water. Quite aside from the impenetrable cold and darkness, water undergoes some odd changes as its depth increases. With every 10 m the pressure goes up by 1 atm, or 14.7 lb per square inch (psi). At a depth of 100 km water bears down with 10,000 times the pressure of air at the surface of Earth. If the gravitational pull is greater, this effect is magnified further. At such high pressure water undergoes a transformation. Instead of flowing like a liquid it crystallizes, forming a material called Ice VI. At somewhat higher pressure this transforms once again to Ice VII.

So what are these odd forms of our most precious liquid? Ice VI has water molecules arranged in a cubic crystal structure, which then breaks down to a less ordered crystal structure at higher pressure, forming Ice VII. Thus the bottom of our ocean would be frozen, even if its temperature was above 0 °C.

Our aquaplanet might be positively balmy at the surface but be thoroughly frozen in its greatest depths. Below this layer of ice, the water would give way to a rocky crust and mantle atop a metallic core, much like Ganymede.

Once again, these worlds seem highly probable and should reveal themselves through the investigative methods described earlier. Within 10 years we should have a catalog of planets of all sorts and sizes. We should also have a robust idea as to which of these support life, or are capable of doing so. Moreover, of the potentially habitable worlds, we might know which are liable to host intelligent life.

Dream Worlds

Leaving the known and possible worlds behind we finally explore the exotica, worlds in which our imaginations soar across impossible landscapes. Worlds made of diamond, pure water or carbon soot are conceivable, even if conditions in the universe conspire to prevent their synthesis.

For example a diamond world, comprised of nearly pure carbon, might seem fanciful, but it is possible. To create such a gem the same sort of stellar evolutionary scenario must pan out that produced the helium world. Swap the helium-white dwarf star for another made of carbon and oxygen. Although the interaction of two carbon-oxygen white dwarfs normally leads to rather explosive fireworks – supernovae – if we could turn off the transfer of material from a low mass carbon-rich white dwarf to its more massive partner, then we might just produce a real gem of a planet.

Within the carbon-rich white dwarf the carbon is structured much like a diamond. As the mass of the white dwarf falls with the loss of its bulk to its partner, the internal carbon skeleton will morph into the structure of a diamond. Such a world would have a diamond interior smothered in a thin atmosphere of carbon monoxide.

What about a water world – a pure water world? Could you build a planet from water alone? Hypothetically yes, but unlike the water worlds of the movies, water doesn't remain fluid under high pressure. Although the planet might be born from nearly pure water, under pressure the compound transforms through a number of phases, forming an ice. This structure develops as pressure forces the water molecules together. Weak bonds between the hydrogen and oxygen atoms that form the molecule cause the molecules to align, forming a crystalline structure.

The interior of a pure water world would consist of a very deep ocean – perhaps over 100 km that solidified with depth. Heat released from the deep interior as the planet formed would be transported by convection within the ice and the overlying water layer. At the very heart of the planet, a small kernel of water-soluble compounds might form a core. These compounds would have come with the water from which the planet amalgamated.

Could such a planet form? It would be highly unlikely, given the sorts of materials that are present in planetary systems. Where would all the other "stuff" that makes planets have gone to? It seems untenable to form a ball of water, free of rock or metal, in a standard, planet-forming scenario. If we ever did find a pure water world, then we'd have to suspect the work of intelligence rather than natural forces.

You might then wonder what would happen if you built a world from porridge or pasta. What would happen then? Oddly, this is relatively easy to work out. Porridge is made primarily of carbohydrate, with a smattering of other carbon-rich compounds, such as proteins. The principle elements are carbon, hydrogen and oxygen in the ratio 1:2:1 – simple stuff. Take an Earth-mass of porridge and let gravity do its stuff and some interesting things happen. The hydrogen and oxygen are heated out of the carbohydrate as the planet coalesces. These naturally form water.

The carbon core of each molecule will then compress until it first forms graphite, then diamond. Our porridge world will develop a diamond core, occupying more than 50 % of its mass. In a thin layer on top of this would be a zone rich in graphite and water, with a deep ocean of water cooked out onto the surface above. If you were lucky the floor of the ocean might still be porridge, but the volume of water would likely be so great that the pressure would compress this into a less palatable layer. The ocean would have a mass greater than the carbon core and thus likely be transformed into a variety of exotic ices below 90 km or so in depth. The interior would be initially warm from the process of accretion and the decay of carbon-14 into nitrogen-14. However, like our diamond world above, the mantle would freeze out and be incapable of any form of prolonged geological activity.

Conclusions

The last two decades have seen a revolution in our understanding of extraterrestrial worlds. Prior to the discovery of the planets orbiting PSR 1257B+12, planetary science was restricted to our own Solar System – and to our wildest imaginings. Now, we have hundreds of confirmed examples of foreign systems of alien worlds. This number will rise into the thousands in just a short time. This vast terrain is a playground for our imagination and our theories. Planets made of iron, helium and water are only a beginning. Although we are unlikely to find a world built from porridge, more exotic abodes may well be commonplace. Watch out for the first interstellar diamond – a planet worth more than its weight in gold.

2. The Formation of Stars and Planets

Introduction

The formation of stars and planets was until recently shrouded in dust, both literally and metaphorically. The basic underlying principle had been understood for over two centuries, having been proposed by the French mathematician Pierre Simon-Laplace and the Dutch philosopher Emmanuel Kant. The proposition was that stars were born from clouds of gas and dust, called nebulae. These beautiful sculptures dot our galaxy and the majority of others. However, such material is opaque to visible radiation.

The advent of infrared and microwave astronomy demonstrated that the nebular hypothesis was correct. However, the nuts and bolts of the mechanism would await higher resolution observations of the guts of these nebulae with long wavelength observations and visible light.

The problem with nebulae is largely down to dirt, or, more precisely, dust. Dust is very effective at scattering visible light, particularly the shorter wavelengths. Images of star fields, in which nebulae lie, show a clear reddening and dimming (or extinction) of the stellar output by these bodies of gas and dust. However, longer wavelengths are able to circumnavigate much of the muck, revealing the viscera underneath. What is remarkable is that observations at long wavelengths beautifully confirm the centuries-old hypotheses; such is the power of the human imagination.

More recently, the sharp eye of the Hubble Space Telescope revealed small protostars forming in the Orion Nebula, surrounded by broiling discs of gas and dust. These "propylids" will in time birth new stars and, perhaps, with some luck, a myriad of small, orbiting worlds.

D.S. Stevenson, *Under a Crimson Sun: Prospects for Life in a Red Dwarf System*, Astronomers' Universe, DOI 10.1007/978-1-4614-8133-1_2, © Springer Science+Business Media New York 2013

Yet, the visionary imaginings of Kant and Laplace are not the final word in planet formation. In recent years it has become apparent that at least some worlds are birthed in alternative and often violent situations. This chapter is the story of planet formation in all its violence and glory. In particular we will examine the precarious origin of planets around red dwarf stars, suggesting why these worlds often bear a dissimilar feel to those formed around more massive stars such as the Sun.

Gravity's Role in Star and Planet Formation

Gravity is the obvious driving force in star formation. It is gravity that turns a cloud of gas and dust, one light year across, into a ball a million or less kilometers wide. After all trillions upon trillions of tons of material is attracting itself and anything else that happens to lie nearby. However, gravity has a battle on its hands. The material may be massive, but it is also diffuse – spread out over trillions of kilometers, with an average density that would make the manufacturer of any earthly vacuum proud. In large part, science fiction dramas are responsible for giving the misleading impression that nebulae are dense. In them ships can hide from one another and light cannot travel more than a few meters. However these imagined fogs are far more opaque than any real galactic cloud. Indeed, one only has to look at photographs of the famous Coalsack Nebula. This particularly dark cloud of dusty gas is translucent enough so that with suitably powerful 'scopes distant stars are visible, reddened but glimmering through the edges of this dark abyss.

These dark chambers of creation are held up against gravity by the internal motions of the gas and by the weak interstellar magnetic field. These forces are all that is necessary to prevent gravity from drawing the material closer together and forming a star. However, give that cloud a nudge, and, if the circumstances are right, the cloud will begin the irreversible and irrevocable process of collapse. The outcome, millions of years later, may be a star and a retinue of worlds.

So what drives the collapse? In most instances it takes more than a gentle nudge to give gravity the edge over the countering forces holding the cloud up. This can be the gentle push of starlight or stellar winds, or something more violent, such as the fiery breath of a supernova, or even the carnage of a galactic collision. Once the density of gas exceeds a critical threshold, gravity can begin crushing the matter inwards.

The next stages are less well understood. But so begins a delicate interplay of forces and energy. The nebula has a very large reserve of gravitational potential energy and so once collapsing will effectively keep on going until something resists it. During the collapse, particles of gas and dust begin to collide and rub together. These frictional forces steadily raise the temperature until material becomes hot enough to emit infrared radiation and later visible light.

During these transformations the bulk of the matter collects into a sizable ball, glowing with the heat of a few thousand degrees; this ball is a protostar. Around the equator of this ball of gas and dust, the remaining matter begins collecting into a swirling disc that steadily feeds the growing protostar. This process is the inevitable consequence of the conservation of angular momentum. As matter streams inwards, subtle variations in the motion of the gas set up a nearly circular motion around the center of gravity. Yet material moving inwards from further afield also has angular momentum. As it moves inwards, like the arms of the proverbial ice skater, the material is compelled to move faster and faster, in a direction roughly parallel to the equator of the star. However, material falling inwards along the rotation axis can fall freely, as it has little or no circular motion relative to the protostellar surface. Thus matter progressively collects into a rotating disc around the protostar.

The protostar is constantly heating up as material streams inwards and friction drains kinetic energy from the mass. The rising temperature sends the protostellar mass across a few thresholds. The first occurs when the temperature exceeds a few thousand degrees Kelvin. At this point hydrogen and later, at higher temperatures, helium ionize. These changes involve the loss of one or more electrons from the formerly neutral atoms of gas. This has two consequences. Firstly the material becomes more opaque to radiation, trapping energy more effectively within the star. Secondly, ionization makes the material electrically conductive, allowing it to respond more strongly to the interstellar magnetic field that permeates the protostar.

The interstellar field initially slips ghost-like through the collapsing gases. However, as the material becomes ionized, the field becomes concentrated within the protostar. One consequence of this is that the star acquires a magnetic field of its own, resulting in some spectacular fireworks. Ionized gases can now launch into interstellar space, flowing along the constricted field lines. Secondly, the developing star can develop complex features such as star spots and flares, as magnetism either resists or assists the movement of energy through the shrinking envelope of the protostar.

Further milestones follow. As the temperature rises over 700,000 K any deuterium, a heavy isotope of hydrogen, can fuse in pairs to form a light isotope of helium – helium-3. At temperatures over 2–3 million Kelvin, any lithium is also consumed. These transient energy sources provide a little additional support and slow the contraction of the protostar. However, it is only once the temperature exceeds 2.5 million Kelvin, under very dense conditions, that the bulk of hydrogen in the core of the star (hydrogen-1) can ignite.

In more massive stars, the density of the stellar interior is lower than the treacly conditions found in red dwarfs. These stars become hot enough, earlier on in their contraction phase, so that hydrogen fuses to make helium at relatively low density. However, a red dwarf collapse is much more exacting and the density of the gases correspondingly higher. Hydrogen fuses sluggishly to make first helium-3 then helium-4 over extended timescales, resulting in low energy outputs and a much lower overall luminosity in red dwarf stars.

If the mass of the protostar is less than 0.075 solar masses (or roughly 1/13th that of the Sun) then the hydrogen never becomes hot enough to fuse to make helium. Instead the object fuses its heavy hydrogen isotope and lithium reserves within 10 million years or so, then slowly cools down, contracting slightly further under gravity. These brown dwarfs may also be home to planets, but ones locked in a diminishing embrace; their central star dimming and fading within a few billion years.

Those bodies with sufficient mass eventually settle down as Jupiter-wide bodies glowing with the light of 1/10,000th that of the Sun. Any planets that there are need to cling on tightly to receive any nurturing warmth. Were the Sun to be replaced by a red dwarf, such as the 10 solar mass Proxima Centauri, even our tightly snuggling Mercury would be an icy ball, struggling to attain temperatures comparable to Jupiter around the Sun.

The planetary systems of red dwarfs more closely resemble Jupiter and its Galilean satellites, than they do our Solar System. Planets tightly circumnavigate their diminutive star in small, fast orbits. These are Solar Systems in miniature.

The Effects of Neighboring Stars

However, the formation of planets around these stars is a tricky business. It isn't that the underlying physics is different. Instead the low mass of the disc makes it highly vulnerable to assault from other, swifter moving stars. A typical massive star makes the journey from nebula to star in less than 500,000 years. In that time the red dwarf hasn't even contemplated its existence. The red dwarf is still a twinkle in the universe's eye.

A billion years later, when all the massive stars have faded to oblivion, obliterating their surroundings first with a barrage of ultraviolet light, then cursing them with supernovae, the red dwarf is still meandering its way lazily down to the point of ignition. Thus the protostar and its incumbent disc of gas and dust will have been thoroughly pummeled by generations of much more massive stars. This can result in significant erosion of material by radiation and winds from other stars, but also mean significant rounds of enrichment by materials shed by these more glamorous objects.

Even the solar nebula, collapsing within 30 million years, shows signs of enrichment by material from supernovae and possibly other intermediate mass stars. There is ample evidence that the solar nebula was showered with radioactive isotopes of aluminum. Their daughter isotopes are found abundantly in certain types of primitive meteorites called chondrites. These can only have been produced in a nearby supernova, as these isotopes have half-lives too short to have been produced earlier in the lifetime of the pre-solar nebula.

Brown Dwarfs

The nearest known brown dwarf is in a distant orbit to the nearby orange K5 dwarf Epsilon Indi, 11.8 light years away. The 40–60 Jupiter-mass brown dwarf is about the same diameter as Jupiter but with a photospheric temperature of ~1,300 K and 0.8–2 billion years old. It orbits the K dwarf distantly at 1,500 A.U., or 1,500 times further from the Sun than Earth does.

At the lowermost end of the HR diagram, and off the bottom of the main sequence, the brown dwarfs grade into gas giant planets, like Jupiter. Indeed, the identification of young, hot, Jupiter-mass objects floating free in the Orion Nebula has created more than its fair share of headaches for theorists and nomenclaturists alike. These young, obscure interstellar hobos, going by the name "orphets" or "planetars" appear to have formed in situ, or they may have been formed orbiting a parent star before being ripped away by a close encounter with another object. Their surfaces have temperatures comparable to class brown dwarfs, and their origins are hotly debated. There are some suggestions that these free-roaming worlds may outnumber stars by several orders of magnitude, dominating the population of objects in the universe.

These orphets, if indeed they formed in situ, will have formed from the direct collapse of small cores of dense gas within a larger nebula. In some instances they may be low mass red dwarf protostars that were robbed of sufficient gas to continue growing to the point of hydrogen ignition. Most are only a handful of Jupiter masses in size and may simply be the starved remains of protostars that continued to contract, heat up, but ultimately failed to ignite their fuel reserves. Some may have simply been bequeathed insufficient mass at birth and ended up with planetary masses. At present, the true scenario is unclear, but they may be very abundant in the galaxy and the dominant type of interstellar body. Stellar purists see these rogue planets as objects that have been stripped from their parental protostars by violent encounters in the birthing chambers. However, some others, approaching the problem from the planetary perspective, are loathe to call them planets and instead suggest that these objects might have formed in situ in the same manner as stars themselves.

Maria Rosa Zapatero Osorio (Instituto de Astrofisica de Canarias) and co-workers have shown that the effective atmospheric temperatures of these super-Jupiter mass objects lie in the range of ~1,700–2,200 K (spectral class L0–L5). Using cooling curves for brown dwarfs provided by Isabelle Baraffe and the known ages of the stellar population, the masses of the objects were derived. These dim red bodies had temperatures matching objects of 5–17 Jupiter masses (S Ori 45 and S Ori 60, respectively), which is at or below the deuterium-burning limit.

Deuterium, an isotope of hydrogen, fuses to form helium-3 at temperatures in excess of 700,000 K. In bodies with masses less than 12 or 13 times that of Jupiter, or less than a sixth that of any red dwarf, deuterium fails to burn, and these objects are considered genuine planets. In some eyes, the lowest mass in which this reaction occurs forms the boundary between a brown dwarf and a planet. Thus, if we hold this definition as being true, many of these wandering worlds are indeed planets and not brown dwarfs. Zapatero considered the possibility that they were all outlying members of the σ-Orions stars system. However, at 660,000 A.U. from σ-Orions, it is not particularly likely that these sub-stellar objects formed in a circumstellar disc surrounding the massive O- and B-class stars comprising the spectacular bulk of this cluster.

Originally, astronomers attempted to define the red dwarfs as stars without lithium in their spectrum and the brown dwarfs as those with lithium. However, given that young stars can retain lithium for a short period after they reach the main sequence, while some brown dwarfs with masses greater than 0.065 solar masses can burn lithium, there exists a very messy divide that may render it useless. The brown dwarfs have caused a problem for those interested in how planets and stars form. Are they really so different? Only when the age of the object is clear can the coolest objects be divided into brown and red dwarfs – or where the mass is known. Although the mass cut-off for hydrogen burning isn't known precisely, it is probably around 0.075 M, with a range from 0.07 to 0.085 depending on the metallicity of the object. The higher the proportion of heavy elements (metals) there are in the star, the lower the mass required to fire up nuclear fusion.

Finally, the coolest brown dwarfs have temperatures that extend from around 1,200 K down to that of around 700 K, with the current record held by the exquisitely titled 2MASS 0415–0935 at 683 K (411 °C). This "failed star" is a paltry 2 millionth as bright as the Sun, which means that in comparison with the galaxy's brightest object it is less than 5 trillionths (1/5,000,000,000,000th) as luminous. T dwarfs are characterized by the presence of methane in their spectra. At these low temperatures these dwarfs are practically invisible, with essentially all of their radiation emitted in the infrared band of the electromagnetic spectrum.

However, the 2MASS survey showed how deficient the best laid plans can be. In the solar neighborhood, 53 new red and brown dwarf stars were identified in the initial survey – stars that had hitherto been too faint to see with conventional optical telescopes. As John Gizis – one of the investigators in the original survey – noted, catalogs of nearby stars were grossly deficient in very low mass dwarfs. These ultra cool M dwarfs are so optically faint that even nearby ones eluded searches based on older sky surveys. The identification of brown dwarfs formed a integral part of the search for so-called MACHOs (Chap. 1). Before 2MASS, and related near-infrared, surveys were carried out, astronomers clearly had little idea about the abundance of the galaxy's faintest stars. It had been assumed that the proportion of stars in the galaxy less than 0.3 solar masses decreased sharply with decreasing mass. 2MASS showed that the proportion of stars with low masses was considerably higher than anticipated. Moreover, the trend in stellar numbers (the mass function) extended into the domain of brown dwarfs. In turn this increase in numbers allowed for a greater proportion of dark matter to consist of faint or failed stars.

The Planets of Red Dwarfs

In general the planets of a red dwarf will hug closer into the gravitational heart of their parent star than those around weightier objects such as the Sun. Moreover, with less mass to birth planets, the worlds that do form will also tend to be lower in mass. Consequently, Jupiter mass companion worlds are rare around red dwarf stars. However, planets with the mass of Neptune or less appear abundant.

One interesting consequence of the sluggish rate of formation of the lowest mass stars is that their planets probably form before they do. This may seem perplexing, given that the Sun was well and truly fired up long before Earth had coalesced from the shards of star formation. However, this is a natural consequence of the more difficult and restricted conditions found around the lowest mass stars. Even within our planetary system the massive planets Jupiter and Saturn may well have beaten the Sun in the race to the finish line.

Rocky worlds, or those with icy cores, appear to take tens of millions of years to form. Jupiter mass worlds may form through a direct process of collapse, perhaps lasting a few hundred thousand years. However, the aggregation of dust to form the first pebbles, then asteroids, then the planetesimal fragments of future rocky and icy worlds takes millions of years. Examination of the dusty discs around nearby stars generally confirms this picture. So, a red dwarf, with its protracted collapse and ignition, may drag on well beyond the stages that form the worlds and conceivably also any life on them. It seems odd to think that life could evolve substantially in the time it takes the central star to settle down, yet this is a consequence of the processes of evolution of life and those of stars – they don't necessarily work in harmony.

Alternative Routes to Rome

Although the process by which planets form seems haphazard and fraught with danger it does appears that all that is needed to form a planet is a disc of material. Indeed, if one accepts some alternative definitions of planets based simply on mass rather than location, it may be possible that planets may form independently of stars.

Planets are born from collapsing discs of material. The question is, how might such a disc form? Although the majority of planet-forming scenarios picture their birthing discs centered around the spiraling heart of a protostar, discs of material can arise in other situations. For example when two stars collide – a process that may be common in dense clusters of stars – a substantial mass of stellar material can be flung violently from the coalescing stellar bodies. Such material should rapidly congregate under

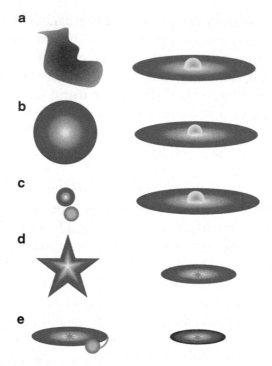

FIG. 2.1 Planets may form in four possible situations. In (a), the Kant-Laplace model, a massive cloud of dust and gas collapses, forming a spinning disc around the central protostar. Planets form within the disc. In (b), a red giant star evolves into a white dwarf, and some of the material from its envelope collects into a disc around the newly formed white dwarf. In (c), two stars collide and merge, but some material is sprayed outwards, forming a disc in which planets can form. In (d), a massive star explodes, leaving a neutron star. Some material from the envelope collects into a disc around the newly formed neutron star. In (e), the neutron star is spun up through an accretion disc of material accreted from a companion. Some of this material collects to form planets around the reborn millisecond pulsar

the influence of gravity into a disc around the resulting star. This allows the process of planet formation to occur around stars that are already mature. Indeed one could envision some form of hybrid planetary system emerging from such a chaotic event. New planets would spawn inside the disc of debris showered outwards in the collision. But there may already have been a planetary system in place. Although such a violet event would most likely displace or expel the original worlds, some could survive in orbital patterns that differ from the new sibling worlds (Fig. 2.1).

Alternatively, as an aging red giant star begins to expel its outer layers in a planetary nebula, some of this material could collect in a disc around the newly exposed white dwarf star. A small number of low mass worlds could begin a protracted life around the steadily cooling ember of their parent star. Similarly, were any of the original retinue of worlds to survive the red giant phase – and examples of this are now known – the resulting planetary system would be a hybrid of old and new worlds with strongly varying histories. There are about a 100 examples of debris discs around white dwarf stars. The Spitzer and WISE 'scopes both have observed countless examples of these since the first was found around the white dwarf Giclas 29–38, in 1987. However these discs are small – typically smaller than the diameter of Saturn's ring system. Thus, at least as the evidence stands at present, the discs that have been observed around white dwarfs seem too flimsy and short-lived to birth anything as sizable as a planet. However, that could well change.

The final, and perhaps the most bizarre, scenario for planet formation around a star owes its origin to the death of massive stars. Aleksander Wolszczan and Dale Frail discovered the first exoplanets in 1992. These tortured worlds were Earth- or Moon-mass bodies, orbiting the millisecond pulsar PSR 1257 + 12. Their presence was given away by the subtle alterations in the twinkling of their pulsar parent, as their orbiting masses tugged on the diminutive star. It is highly unlikely that these worlds would have formed around the parent star in its youth. Massive stars live for inconsequential periods of time – too little for planets to form. Moreover, the orbits of these worlds would easily place them inside the envelope of the star as it ballooned into a red supergiant even before its final demise in a supernova.

However, millisecond pulsars are not quite as they seem. Such pulsars are spun up through the accretion of mass and angular momentum from a companion star. Typically, as the pulsar speeds up it sends out an increasingly energetic field of radiation and stellar winds that ultimately decimates the unwilling donor companion star. With the inevitable decline of the companion, the supply of accretable material dwindles. This may leave behind a retinue of debris from which the planets assemble. Thus these pulsar planets may instead have formed from the remaining debris of the doomed companion rather than from the leftovers of supernovae.

Bizarre and unexpected though pulsar planets are, there are numerous examples of stripped companions around them. Some are reduced to the mass of brown dwarfs, while a few cling onto some semblance of normality as more massive white dwarfs. The oldest known planet in the galaxy orbits a white dwarf and millisecond pulsar pair in the globular cluster M4 (Chap. 2). Its age is estimated at 12 billion years or so, given the age of the cluster.

Energy and Life

Energy is a prerequisite for life. Energy is needed to force chemicals together to form new compounds. It is also required to transport substances within organisms. Life is a delicate balance between chemical synthesis and the drag of entropy. Energy repairs and builds organisms from simple molecular building blocks. Perhaps most significantly, all organic life on Earth derives from two processes: photosynthesis and, less obviously, chemosynthesis in bacteria living in more extreme and dark environments.

Photosynthesis is a seemingly ubiquitous process, driving the development of the majority of multi-cellular life on the planet. In these series of interwoven reactions, carbon dioxide is fixed; becoming chemically "reduced" and combined with other organic compounds in a series of reactions that ultimately produce other essential organic molecules, such as amino acids (Chap. 6). Reduction is a process by which substances gain electrons – commonly as hydrogen – and become more reactive, more energetic. Glucose is the reduced end-product of photosynthesis, but the starting product for the rest of organic chemistry in higher plants and the animals that feed upon them.

Glamorous though the world of plants is, it is easy to forget that the basic processes of photosynthesis began in simpler bacterial forms that are still found in abundance today. It is even easier to overlook the vast microbial world that derives its energy source from the geothermal reserves within our planet – and conceivably on other worlds as well. This is chemosynthesis. One of the earliest examples of this extremophile lifestyle was found thousands of meters down in the Columbia River basalts – a seemingly unforgiving mass of 17-million-year-old iron-rich basaltic volcanic rocks.

Fed only by the water percolating through the rock pile, ions dissolved from the rocks, along with carbon dioxide dissolved in the water, and colonies of hardy microbes were able to convert the carbon dioxide into food molecules. The underlying chemistry was based on three components: iron, water and carbon dioxide. Geothermal energy was the final ingredient in this extraordinary mix, providing the necessary kick to initiate and sustain the life-giving reactions.

The advent of deep-sea exploration revealed many more peculiar organisms living in dark, forbidding environments. Around undersea geothermal vents life clings precariously but abundantly to the broiling mass of mineral-rich waters pouring out of the vents. So efficient are bacteria at obtaining life from this brew that far more complex animals live alongside them, feeding on the bacteria as food or using these bacteria internally to produce food directly. Heat neatly substitutes for sunlight. Biology isn't fussy what the energy source is. As long as one is present, life apparently finds a way. The discovery of this hidden biosphere not only confirmed Thomas Gold's contention that there existed a hot, deep biosphere on Earth, but opened the door to the existence of extraterrestrial life in some extreme environments, many that put the concept of the stellar habitable zone in the shade.

Thus energy is the key ingredient for all else that happens on a planet. Get the balance of energy input and loss correct, and planetary chemistry can evolve into biochemistry – life takes hold. What then are the energy sources available for young planets?

Planetary Heat

Aside from radiant energy from its parent star, the majority of energy of energy inside a planet is bequeathed to it at birth. There are various potential sources of energy that contribute to the total available budget of a planet.

At formation two principle energy sources come into play. The first is the energy of accretion. As planetesimals collide and fuse together their kinetic energy is converted into heat. Rough estimates suggest that this alone should be enough to melt at least the outer portions of planetesimals, allowing their fused remains to collect into a largely molten ball of rock and metal. As the mass of the planetesimal goes up, however, most of the heating will be confined to the upper terrains of the growing world. Increasing compression, as well as more restricted heating by impacts, should allow even a thoroughly battered world to at least partially solidify early on. Impacts tend to radiate most of their energy into the cold vacuum of space, rather than conducting it through the insulating mass of rock in the planetesimal. For instance, although there is ample evidence that the Moon was bathed in a magma ocean, this was a relatively superficial feature confined to the upper portion of its mantle. The lower mantle was at best a grainy mush, rather than a liquid.

Perhaps the greatest contributor to planetary heating is radioactivity. In essence this comes in two flavors – although the distinction is clearly blurred. The majority of radioisotopes in nature have relatively short half-lives – the time it takes half the mass of radioactive element to decay. These short half-lives mean that the energy released when these elements decay is only able to participate in the heating of the earliest stages of planetary formation. Moreover, these isotopes primarily come from supernovae – specifically core-collapse supernovae, produced by the deaths of massive stars. Therefore, if they are to form a large part of the early history of a planet, it has to have formed in a tough environment near one or more massive stars that subsequently blew themselves to bits. The star system has to be close enough that it is showered with the necessary isotopes, but obviously not so close that the infant solar system, and its retinue of accreting planetesimals, is obliterated.

Principle among the short-lived isotopes is aluminum-26. This isotope of aluminum has a half-life of 770,000 years. Every 770,000 years the amount of aluminum-26 remaining decreases by half its original amount, and thus the amount of energy available to heat the interior is short-lived. In terms of planet formation a million years is a short period of time. Yet, the asteroids of our Solar System contain ample evidence of inclusion of and heating by aluminum-26. Thus, unlikely and presumably rare though such a scenario sounds, it does appear that at least our planetary system was fortuitously placed to be blitzed and enriched by a nearby supernova.

Long-lived radioisotopes may well be the biggest contributor to planetary warmth. Although most of these elements have an origin in core-collapse supernovae, others are formed, less abundantly, through more leisurely processes occurring in red giant stars. These long-lived isotopes are able to heat the interior of planetary bodies for billions of years. Most important among these are uranium, potassium and thorium. Uranium-238, in particular has a half-life as long as Earth is old, roughly 4.5 billion years. Earth may be a relative senior, but its mantle and crust are still relatively rich in this element. Far more abundant, but with a shorter half-life, potassium-40, decays through three different routes into argon or calcium, with a half-life of 1,260 million years. Potassium is an abundant element and thus has a very significant heating effect on planetary bodies. Even if a planet is solid at inception, heating by long-lived radioisotopes will soon raise the temperature of their interiors to the point at which they can convect.

Many isolated low-mass nebulae – the Bok globules – appear far removed from the site of the formation of massive stars. Thus these "stars-in-waiting" may never experience both the shock of a supernova nor its chemical enrichment. Since many red dwarfs are likely to be formed from these low mass nebulae, they, too, will likely be deprived of these energetic elements.

What else can heat a planet up? There are three further sources that are known, plus a fourth that is possible if not proven. The first is simple compression. This prosaic method of heating is easy to demonstrate using a bicycle pump and a bit of muscle power. Putting one finger over the air pipe and pumping the piston with the other soon heats the air up inside the pump. Although the precise physics of compressing rock is somewhat different from gases, the principle is the same. Squeeze a rock enough and it will heat up. All planetesimals will heat by this method once their diameter exceeds a few hundred kilometers.

Most rocky worlds are endowed with a significant amount of iron. Iron is produced copiously by thermonuclear (Type Ia) supernovae and to a lesser extent by core-collapse supernovae. Thermonuclear supernovae obliterate white dwarf stars if their masses exceed a critical limit. These eruptions are so common and so prolific in the creation of iron that this element is one of the most abundant heavy elements in the universe, after oxygen and carbon. The interesting thing about iron is that it's dense. Once enough of it collects in a molten mass, it will descend by gravity to the core of the planet it finds itself in. Indeed, my namesake, Professor David Stevenson of Caltech, rather famously suggested that we could send a probe to Earth's core by imbedding it in a large mass of molten iron. Simply pour this mass onto a suitable fissure and let gravity do the work. The molten iron will neatly burrow its way through the crust of Earth, and once into the mantle will descend with relative speed to the core of the planet, taking the probe with it. Most importantly, as this mass descends, its gravitational potential energy is converted into heat, ensuring that it doesn't freeze on the way down. Perhaps because of the current tough economic times, no one has taken Professor Stevenson up on this suggestion, reluctant to send such an expensive mass of iron into the planet's interior...

A planet with a rocky composition is often referred to as chondritic. This refers to the type of meteorite found in our Solar System with a composition thought to be derived from the original solar nebula. These meteoritic fossils from our planet's formation indicate that our solar nebula had a composition enriched by both supernovae and ancient red giant stars. The bulk Earth has a composition that is broadly chondritic in nature. However, recent work shows that the bulk Earth deviates somewhat from this chondritic composition. The explanation for this discrepancy is probably a result of Earth forming through the collision of large planetesimals that had already differentiated into a core and mantle, thus altering the proportions of some key elements. There is ample evidence, for example, that the Moon was formed late on through one such collision (Fig. 2.2).

In the early Earth the quantity of iron was sufficient to endow our young world with a large iron core. The formation of the core would have released more than enough energy to melt the bulk of the planet. However, this impression is somewhat misleading, as the process of core formation, although rapid on a planetary timescale, was undoubtedly haphazard. With each impact a little more iron was liberated, trickling down into the interior, releasing its store of gravitational energy. Much of the iron would have arrived as iron oxide, which had to be brewed appropriately before the iron contained was liberated. Thus, although the

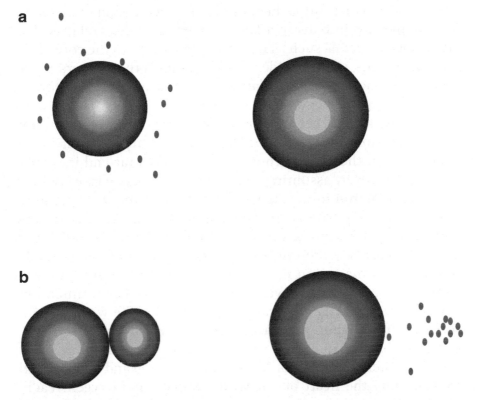

Fɪɢ. 2.2 Two models for the formation of terrestrial planets such as Earth. In the simplest model, (a), the planet accretes from dust grains, rocks, etc., that have the same overall composition as the original nebula in its vicinity. The core then segregates from this material. This "chondritic model" assumes that Earth's bulk composition is equivalent to this. However, analysis of various elements and isotopes shows that Earth's composition is non-chondritic – it differs from the composition of the chondritic meteorites in the Solar System. Instead the composition is best explained if Earth formed from planetesimals that had already differentiated into a core, mantle and crust (b). During the collision of these large bodies, some lighter elements can be lost to space, while the heavier iron-loving elements are enriched. Most models for terrestrial planet formation favor this scenario, with multiple late collisions forming Earth and later the Moon

back-of-the-envelope calculations suggest that Earth was molten early on, it may not have been, particularly if the heat released by the various processes that formed it were sufficiently protracted to allow heat to escape at a more leisurely pace.

Once a planet forms, heat drives the segregation of elements and compounds into distinct layers, dictated by their relative densities. Dense metals such as iron make their way to the core of the growing worlds, while less dense silicates form thick layers above. Once cool enough, more volatile gases and liquids then form a thin, low-mass veneer on top.

In general, a planet orbiting a young star should start out hot. The proportion of radioactive elements will undoubtedly vary, depending on the neighborhood in which the star and its planet formed. However, assuming that it has been bequeathed a similar fraction of iron to the planets forming the inner Solar System, it should be able to melt its interior as the iron core coalesces. Any world with a mass similar to Mars or Mercury will retain at least a partially molten iron core for billions of years. Rocky planets with masses in excess of Earth should retain a molten iron core for 5–10 billion years, at least, and possibly much longer. This assumes a chemical composition for the world similar to Earth. If, for whatever reason, an alien world was born with a composition rich in carbon things could be very different.

The interior of Earth is dominated by silicon and oxygen, with iron forming the heart of our world. Silicon and oxygen readily combine to form the ubiquitous silicates of our rocks. At modestly high temperatures such rocks either melt or deform and can thus convect inside our planet. However, carbon in its reduced form will form diamond under high pressure and temperature. On Earth diamonds are rare, reflecting their limited rate of formation inside our planet. But if a planet had a composition much richer in carbon, diamond might come to dominate its structure. Not only is diamond hard, it has a high melting temperature. Any planet rich in carbon would find its mantle solidified into a priceless but rather inaccessible crown jewel.

Why is this important? Well, mantle convection is important for life. Planet-wide convection, driven by the slow release of internal heat, drives two apparently essential life-giving processes. Mantle convection allows a planet to recycle its crust, its water and its carbon dioxide. Without mantle convection, carbon dioxide will steadily deplete in the planetary atmosphere, as it reacts first with any water, forming hydrogen carbonate, then with calcium and other ions to form solid calcium or magnesium carbonate rock

(Chap. 5). Since plants and other organisms rely on carbon dioxide gas as a building block for organic compounds, its loss would be catastrophic. Mantle convection ensures that carbonate rocks eventually meet their maker and are converted back into silicates and carbon dioxide gas in the hot interior.

Mantle convection also ensures that the outer core of our planet, and presumably other planets, is continually chilled from the outside. This, in turn, drives convection in the core. Coupled to planetary spin, convection drives our geodynamo – a strong planetary magnetic field – that protects the planet's atmosphere from erosion by the solar wind. Without a geodynamo, Earth's atmosphere would be steadily eroded by charged particles and radiation coming from the Sun. It would appear as though this, too, would be important to an alien world.

There are two contradictory caveats here. On the one hand Venus, deprived of an ocean to cool its surface, appears to lack plate tectonics and a strong planetary magnetic field – despite its implicit molten iron core. And although there is abundant evidence of atmospheric erosion by the solar wind, Venus still retains a very significant atmosphere after billions of years of abuse. Thus a planetary magnetic field may be compensated by other effects. In the case of Venus this is likely to come down to abundant, if periodic, volcanic activity and an inherently thick atmosphere caused by a dry surface that was unable to absorb carbon dioxide.

Any habitable world orbiting a red dwarf has to nuzzle into its host star. To retain enough heat from its ember of a parent, a red dwarf world must orbit tightly within a handful of millions of kilometers from the stellar surface. Such proximity means that such a world will be tidally locked to its parent star. Unable to rotate relative to the stellar surface, the planet merely spins on its axis once for every revolution of its star. Although the year on such a world will be short by our standards, it may still be too long to thoroughly stir its metal core. Thus a red dwarf world may be hot and iron-rich inside, but still lack an appreciable magnetic field.

More interesting still will be the effect of the close proximity of the world and its star on the stellar magnetic field. Embedded deep within the gravitational potential well of its star, the habitable red dwarf world will also lie deep within the star's magnetic bubble. Such a world will not only be buffeted by any outbursts from its star, but may be magnetically coupled to it via its iron core. Aurorae could be a permanent feature of this world, its atmosphere entwined with that of its parent star. This might also lead to enhanced rates of atmospheric erosion by the stellar wind.

There is one final source of heat available to red dwarf planets, depending on the nature of its orbit and the presence or absence of other massive planets close by – tidal heating. The effect of tides can be seen on the Jovian moon system. Io is the closest large moon to Jupiter. Orbiting deep within Jupiter's gravitational well it orbits with its surface tidally locked to its giant planet host. However, as it orbits Jupiter, it is pulled to and fro by the other large Galilean satellites. These causes Io to wobble in its orbit, resulting in large tides that deform the surface by 75 m each revolution. This kneading causes Io's interior to flex and bend, which in turn causes a large amount of tidal heating. As a result Io is thoroughly melted at least within the top 100 km or so of its bulk. The result is massive vulcanism on Io. Similarly, for at least some of its time in orbit around a red dwarf, any rocky world will be similarly flexed and heated, particularly if it has any massive planetary neighbors.

As a consequence of tidal heating Io is effectively boiled dry, suffering the strongest heating in its violent game of tug-of-war with Jupiter and the other Galilean moons. Further out is Europa, cracked and its water-rich surface melted and bruised, while the more distant and more massive Ganymede retains enough heat to maintain a molten iron-rich core.

A red dwarf planetary system is not unlike that of Jupiter and its moons. Swap Jupiter for a red dwarf and scale up the mass and diameter of the moons, and you'd have some idea of the typical red dwarf planetary system. Tidal heating of these worlds must, therefore, be significant, particularly early on when the planets are young, still spinning rapidly and suffering changes to their orbits through collisions and other processes. Some researchers have suggested that this process may be sufficient to effectively boil red dwarf worlds dry. If this were true they would constitute very poor hosts for life. However, tidally-driven dessication may e averted by later impacts (Chap. 10).

On worlds with large gaseous or liquid outer layers, heating will be dissipated in enormous tides, rather than in the heating of the rocky interior. However, in principle the orbits of any red dwarf worlds may experience significant tidal heating for at least the earliest fraction of their lifetimes.

Atmosphere, Hydrosphere and Biosphere

Our world has formed. Its surface is wracked by volcanic activity, its mantle vigorously convecting and its rotation steadily slowing as tidal forces between its mantle and its parent star brakes its spin. After some significant maneuvering, tidal forces eventually push the planet outwards, perhaps disturbing more distant siblings in the process. Within a few hundred million years of inception the planet and its siblings have settled into stable, roughly circular orbits, nestled around the parent star.

When planets form their masses change relatively rapidly. This changes their gravitational influence on one another – as well as on their parent star. Consequently, their orbits tend to migrate in response. Moreover, left over from the spoils of planetary creation are billions of smaller shards of rock and ice. All of this material gravitationally interacts with itself and the growing planets. A key ingredient in this rich brew is angular momentum. Recall the effect of angular momentum on the parent star. As matter spirals inwards the speed of its rotation accelerates. Similarly, any planet moving through a debris disc around its parent star will exchange angular momentum with the material. Material accreting from the outside will tend to cause the planet to gain angular momentum and move outwards, while material accreted from within its orbit will cause the planet to move towards its parent star, through the loss of angular momentum. The balance of these two competing effects will determine the final resting place of any forming planet.

Some unlucky planets may spiral all the way into their parent star. Indeed such a migration is suggested by hot Jupiter worlds, massive and presumably gaseous planets found uncomfortably close to their parent stars. Similarly there is evidence for extensive planetary migration in our Solar System, while in its infancy. Jupiter appears to have migrated close in to where Mars is currently before retreating back out to its present location. Uranus and Neptune may have started much closer in, before being flung outwards by the migrating bulk of Jupiter. As the giants of our system moved, they would scatter lesser icy worlds inwards, delivering water and other volatile compounds to the inner Solar System. The possibility and effect of planetary migration is discussed more fully in chapter 10, in relation to the habitability of Gliese 581d and Gliese 667Cc.

At birth, or shortly thereafter, when the Moon was formed, planets would have been strongly heated and probably fairly dry. However, we are a relatively wet world now, with enough water to cover three quarters of Earth's surface. Most likely this water arrived soon after birth, when the giants shuffled around the Solar System, scattering lesser bodies inwards, or outwards into the depths of the outer Solar System.

For a red dwarf system, there is no reason to suggest that similar dramatic events would not happen there as well. Thus even though red dwarf planets are born deep in the potential well of their parent star and are likely heated in the extreme by differing forces, they should also receive enough volatile material to form oceans or a thick atmosphere.

Thus although it is possible that red dwarf worlds are born dry, there is no reason to believe that they stay that way. Like on Earth, water would arrive later on, assuming that the outermost regions of the protoplanetary disc remain cold enough for ices to re-sublimate.

Examination of ancient, microscopic crystals of zircon indicates that Earth probably had oceans by 4.2 billion years ago, or less than 400 million years after birth. These crystals tell a tale of an Earth with at least some granitic rocks at this stage. Granite is the building block of continents and only forms in the presence of liquid water (Chap. 4). Deep inside Earth, water sweats granite out of the hot mantle rock. As far as we are aware, granites only form this way. Thus these zircon crystals tell a tale of a planet that was cool enough, very early on, to allow liquid water to permeate its surface and crust.

The Moon was formed from a collision between the infant Earth and a Mars-sized planetesimal only about 400 million years ago. Thus, although Earth was undoubtedly hot at birth, it had quickly acquired enough icy material and was cool enough to form oceans.

Conclusions

Perhaps 400 million years or so after birth, the briefest twinkle in the eye of its low mass star, our red dwarf worlds could well have a surface bathed in water, nourished by a thick atmosphere. Their parent star is still contracting and heating up. Violent stellar winds and outbursts are buffeting the planetary atmosphere, while much of the remaining planetary debris is crashing onto its surface – but for now it is safe.

Red dwarf stars undergo such protracted births that they are effectively born after the planets that circle them. Although this leads to some unique circumstances, the overall pattern of planet formation remains similar to that demonstrated by planetary systems like ours. We should, however, be mindful that the worlds of a crimson sun are born in a different environment from those surrounding Sol – and this, in turn, may lead to some unique developmental features that we are unaccustomed to. Only future exploratory missions will be able to determine whether this is true.

3. Stellar Evolution Near the Bottom of the Main Sequence

Introduction

A star is a factory with a fixed supply of raw materials. The nature of its life is dictated by the amount of mass, and to a lesser extent, the nature of the materials bequeathed at birth. In essence the lower the mass, the cooler the star and hence the more sedately the star burns its fuel. This book is dedicated to the coolest stars in the universe – the orange, K-class and more particularly the red, M-class dwarf stars.

The development of life is a slow and perhaps inevitable process whereby organic chemicals become organized in such a way that they are able to reproduce and evolve. Such steps are intrinsically improbable and therefore require sufficient time in which to occur. Stellar lifetimes measured in millions of years provide insufficient time to form stable planets, ruling out massive stars as worlds that might host life. Similarly, less massive stars, those with masses a handful of times that of the Sun, will only live for hundreds of millions of years at most. Almost certainly this, too, is insufficient for anything more complex than simple unicellular existence. Low mass stars, by contrast, have lives spanning billions of years or more. In terms of longevity the M-class dwarfs put all other stars to shame – expiring after hundreds of billions or even trillions of years – far longer than the current age of the universe.

The prolonged and largely stable lives of these two classes of stars make them the universe's most hospitable locations for life-bearing worlds. However, that claim comes with a few caveats that are explored in later chapters.

D.S. Stevenson, *Under a Crimson Sun: Prospects for Life in a Red Dwarf System*, Astronomers' Universe, DOI 10.1007/978-1-4614-8133-1_3, © Springer Science+Business Media New York 2013

Stellar Evolution and the HR Diagram

Stars that burn hydrogen occupy a location on a chart of surface temperature and luminosity. This is the eponymous Hertzsprung-Russell diagram. Of interest here is the long diagonal band of stars that meanders from the top left to bottom right of the diagram (Fig. 3.1). This is the main sequence and contains the 90-odd percent of stars in the universe that are converting hydrogen to helium in their stellar cores. The point at which a newly formed star is located on the main sequence is called the Zero Age Main Sequence, or ZAMS for short.

The red dwarfs occupy the lower third, right-hand corner of the main sequence. They are a diverse bunch of objects that range in mass from a little over half that of the Sun down to around a thirteenth (0.5–0.075 M). The nearest star to the Solar System, at 4.1 light years, is Proxima Centauri – spectral class M6, the

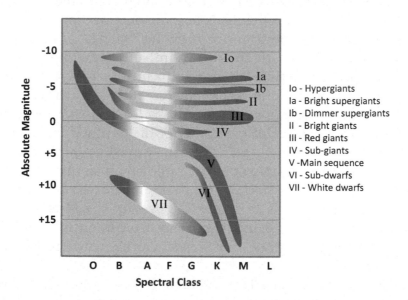

FIG. 3.1 The HR diagram in a more modern guise showing luminosity classes I through to VIII: *Io* hypergiants, *Ia* bright supergiants, *Ib* dimmer supergiants, *II* bright giants, *III* red giants, *IV* sub-giants, *V* main sequence, *VI* sub-dwarfs, *VII* hite dwarfs. As well as inclusion of terms such as hypergiant and sub-dwarf, spectral class L (showing lithium) have been added. Class L is primarily populated by brown dwarfs ("failed stars"), although a few bona fide hydrogen-burning red dwarfs lie at the brighter end of class L

outermost object in the Alpha Centauri triple star system. Despite its proximity, this 0.1 solar mass star cannot be seen with the naked eye. In comparison to the other two stars that make up this stellar triplet, this star is practically invisible, with approximately 1/10,000th of Alpha Centauri A's energy output.

To get an idea just how faint the red dwarf stars truly are, for the *brightest* red dwarf to be visible to the naked eye, the star would have to lie within 1.3 light years of the Sun. In stark contrast, the brightest and very massive O-class stars – although very rare – can be seen over intergalactic distances.

The red dwarfs occupy spectral class M and the hottest portion of the recently added spectral class L. In terms of classification any star on the hydrogen-burning main sequence has the Roman numeral "V" added to its spectral class. Lastly, a second numeral is added to help signify the strength of certain spectral features – usually the absorption of light by one or more chemical elements. The lower the number the hotter the star is within the spectral class. This gives a star's spectral classification three characters. For example Barnard's Star, a red dwarf with 6/10th the mass of the Sun, has a classification of M1V. Meanwhile the cooler and less massive Proxima Centauri has a classification of M6V.

Spectral Features of M and K Dwarfs

The vast majority of red dwarfs occupy class M on the main sequence of the Hertzsprung-Russell (HR) diagram. However, since around the turn of the millennium things have gotten a little more complex on the nomenclature front with new stellar and sub-stellar classes having been added to the lower corner of the diagram. A small but sizable proportion of red dwarfs occupy the brightest portion of the more recently added class L on the HR diagram. This new stellar class was formerly reserved for the brown dwarf stars (Chap. 2). These faint red dwarf stars are in the lowest possible stellar range of 0.08–0.075 solar masses. And just to add a little more complexity, a few very dim M-class dwarf stars are merely young brown dwarf stars that are still in the process of cooling into oblivion. These young M6 to M8 dwarf stars will simply cool and fade, transiting class L before fading into darkness.

Spectral class G stars, like the Sun, have surface temperatures in the range 5,000–6,000 K, while cooler K-class stars have temperatures nearer 3,900–5,000 K. The coolest and dimmest stars are the M-class or red dwarf stars with temperatures less than 3,900 K, down to around 2,200 K at spectral class M8. However, the improved spectroscopy and higher magnification of the 2MASS and other more recent faint object surveys have marginally extended the region of M-class stars down to a dismal M8.5 (2,250–2,000 K). Analysis then extended the domain of the red dwarfs into class L. The M-class stars lack lithium in their spectra, having consumed it in the nuclear reactions that drive convection throughout their bulk. The L-class objects may retain some for a time, hence the L label of the group, but ultimately this, too is consumed in the more massive objects as convection drags it from the outside of the star to the hotter interior.

Finally, the T-dwarfs (brown dwarfs) have temperatures that extend from around 1,200 K down to that of around 700 K, with the current record held by the exquisitely titled 2MASS 0415–0935 at 683 K (411 °C). This failed star is only two millionth as bright as the Sun, which means that in comparison with the galaxy's brightest object it is less than 5 trillionths, or 1/5,000,000,000,000th, as luminous. T dwarfs are characterized by the presence of methane in their spectra. At these low temperatures these dwarfs are practically invisible, with essentially all of their radiation emitted in the infrared band of the electromagnetic spectrum.

The spectra of M-dwarfs are exceedingly complex. As the temperature falls below 5,000 K, hardy molecules such as carbon monoxide (CO) and hydroxyl (OH) emerge from reactions between carbon, hydrogen and oxygen in the stellar atmosphere and photosphere. At even lower temperatures metal oxides, various organic molecules such as cyanide (CN) and compounds of carbon and hydrogen (CH) also appear in the stellar spectra. Descending further, below ~2,500 K water (H_2O) appears, and finally below about 1,300 K methane (CH_4) emerges in the spectra as further reactions between hydrogen and carbon monoxide produce water. The presence of methane in the spectra of T-dwarfs is indicative of very low temperatures (along with the lithium found in the more massive dwarfs).

At temperatures exceeding 1,200 K methane reacts with any oxygen present (predominantly as OH, H_2O or metal oxides). This explains its absence in hotter objects. Therefore, in M-class stars carbon is thus found predominantly as carbon monoxide, rather than as methane.

All of this chemical broth blocks a substantial portion of the energy produced in the stellar core. Below 5,000 K the opacity of the stellar substance, or how easily light can escape its material, decreases rapidly to a minimum near 2,500 K. Below this point, the material becomes very dense to light and relatively little escapes. In part this is due to the condensation of grains and fluff from the cool stellar broth. This high opacity explains why the dirtiest stars, those with the greatest proportion of metals, can fire up their engines at decreasing masses, while metal-poor stars have to be more massive before hydrogen ignites (Chap. 11). Metals form compounds that trap energy and keep the stellar interior warmer at any given mass. They also ensure that the coolest stars emit a greater than predicted amount of radiation in the infrared rather than at visible wavelengths.

As mentioned earlier, astronomers at one time attempted to define the red dwarfs as stars without lithium in their spectrum and the brown dwarfs as those with lithium. However, given that young stars can retain lithium for a short period after they reach the main sequence, while some brown dwarfs with masses greater than 0.065 solar masses can burn lithium, there exists a very messy divide that may render it useless. The brown dwarfs have also caused perturbation for those interested in how planets and stars form. It was assumed that brown dwarfs formed in situations more analogous to planets than to stars. However, are they really so different?

Evidence has emerged supporting the notion that brown dwarfs form in the same way as stars. It appears that, for whatever reason, they ran out of building materials early on and ended up small. Therefore, instead of an underlying mechanistic difference, it appears the only real way to tell a young brown dwarf from a red dwarf is to wait until it ages. Alternatively, we can hope that it is part of a binary system where the mass of each can be accurately determined using observation and a bit of mathematical footwork. Only when the age of the object is clear can the coolest objects be divided into brown and red dwarfs – or when the mass is known. Although the mass cut off for hydrogen burning isn't known precisely, it is probably around 0.075 M, with a range from 0.07 to 0.085, depending on the metallicity of the object.

The 2MASS survey (Chap. 2) clarified the stellar to sub-stellar boundary. Prior to 2MASS very few brown dwarf candidates were known, and without a sufficiently large population of objects to analyze it wasn't at all clear where the division between true stars and sub-stellar objects lay within the main sequence. The brown dwarfs found prior to this time were all much cooler than a typical M-class dwarf. However, 2MASS identified hundreds of faint objects with spectra ranging from M5 to T. Of these objects there was no direct way to assess their age. However, based on their orbits around the galactic core, the faint red dwarfs could be divided into two groups of stars with high or low velocities, relative to the Sun. The former belonged to the galactic halo, the ancient sphere of scattered Population II stars and globular clusters, whereas the latter belong to the main galactic disc.

Stars in the halo Population (II) have orbits around the galactic center that are relatively chaotic and take them plunging through the galactic disc. Disc stars, by contrast, have relatively sedate orbits that broadly parallel the orbit of the Sun around the galactic nucleus. This behavioral difference allowed the ages of each group of dwarfs to be determined. As such late M-class stars, belonging to the disc Population, could be young brown dwarfs that are still heated by a combination of gravitational contraction and deuterium fusion reactions and, in the case of the more massive brown dwarfs (0.065–0.075 M), lithium. Alternatively, they could be older, very low mass red dwarfs with stable hydrogen burning. Similarly, early L-class stars could be young, lower mass brown dwarfs or older low mass red dwarfs. The best that could be said was that early L- and late M-class stars belonging to the high velocity halo Population were likely old and hence to still be shining must be true, hydrogen-burning stars.

Determining the overall metallicity of these dwarfs becomes a useful factor in our attempts to sort out this morass. As Population II stars are older they are also less well endowed with metals. Spectroscopic analysis of these borderline (late M to early L) dwarfs will help discriminate between truly old hydrogen burning stars and those young brown dwarfs. Thus, although we are only beginning our journey up the main sequence already the situation is complex and often contentious.

Spectral Features of K-Class Stars

At higher temperatures absorption lines due to metals and complex organics fade away, while the Balmer absorption line of hydrogen strengthens. At these higher temperatures metal atoms tend to lose electrons more readily and become more strongly ionized. Moreover, the chemical compounds in which they are found at low temperatures are thoroughly vaporized and tend to decompose. K-class stars thus display a more bland but easy to interpret pattern of chemistry than their cooler M-class cousin. With less absorption of visible radiation and a higher temperature, more of the energy emitted from K-class stars is in the visible range of the electromagnetic spectrum. Assuming that the atmosphere of a habitable world was similar to ours, the daytime sky of a planet orbiting a K-class star would probably be as blue as ours, or nearly so. Sunsets might well be more spectacular, with more red and orange light and less blue, but the daytime sky wouldn't be orange.

Structure of M- and K-Class Stars

Red dwarf stars are uniquely structured objects that helps explain their efficiency and longevity. More massive K-class dwarf stars, and those like the Sun, have distinct layers where energy is transported by radiation and by convection. In the inner portions of stars with masses exceeding roughly one quarter to one third that of the Sun, energy is transported by radiation – the movement of particles of light from electron to electron within the hot dense plasma of the stellar interior. Further out from the center, where conditions are cooler and less dense, radiation gives way to the motion of gas. Hot gas rises upwards and cools before sinking back into the interior to heat up once more. This motion – convection – is restricted to the outer layers and is thus only able to mix these layers (Fig. 3.2).

FIG. 3.2 The structure of a red dwarf star. Energy is transported through-out the star by convection, the roiling motion of gas in the thick, treacly interior of the red dwarf. Energy finally escapes as radiation at the stellar surface, or photosphere

A red dwarf, by contrast, with a mass less than one third that of the Sun, has convection throughout its bulk for most of the duration of its life. Thus fresh hydrogen is continually supplied to the nuclear furnace, while the helium ash is mixed outwards. The star as a whole slowly transforms from an object rich in hydrogen to one rich in helium, as nuclear reactions progress. Convection ensures that the bulk of the available stellar fuel is used. A star like the Sun is only able to draw on the material within its core, restricting its lifespan (Fig. 3.3).

The K-class dwarf stars are structured like the Sun. Convection occupies perhaps half the stellar girth, with the proportion of any star affected by the process and continually shrinking as the stellar mass is increased towards class G. These differences in stellar struc-ture have profound effects on the later evolution of these stars.

The Stellar Furnace

Stars of less than 0.25–0.3 solar masses are fully convective (Fig. 3.2), meaning that the entire stellar composition is kept homo-geneous throughout their main sequence lifespan. Helium, minted in the core, is continually removed and dispersed throughout the

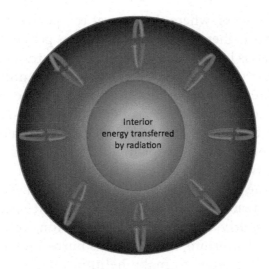

FIG. 3.3 The structure of a K-class orange dwarf star such as Epsilon Eridanae. Convection carries heat only through the outer 50–75 % or so of the stellar structure to the photosphere. The proportion diminishes as the mass of the star increases. Radiation carries energy through the heart of the star

star as a whole, while fresh hydrogen is drawn inwards, mixed and eventually burned. This property makes the lowest mass red dwarf stars unique in the stellar zoo. All objects more massive than 0.25–0.3 solar masses retain a separate envelope that only changes in composition after the star has left the main sequence – and even then in a less dramatic fashion.

As stellar mass increases to that of the Sun, the proportion of the star that convects progressively decreases. The Sun's outermost 25 % convects, but convection eventually ceases when the mass of the star reaches ~1.5 solar masses (around F5V). However, at masses larger than this, core convection – driven by the more energy efficient CNO cycle – commences. There is some evidence that stars in the range of 1.5–1.9 solar masses retain a thin convective skin, and some of these, residing in the instability strip, may have a convective layer lying immediately under their photosphere driven by the ionization and recombination of helium.

Core temperature at Zero-Age-Main-Sequence (ZAMS) starts at around 2.5 million K for the smallest 0.075 solar mass dwarfs and rise to approximately 7 million K for stars of 0.5 solar masses (class M0V). Hydrogen normally burns at densities of ~150 grams per cubic centimeter (or about 150 times the density of tap water).

In the smallest stars, where the central temperature is lowest, the densities are far higher (up to 1 million grams per cubic centimeter). Where the mass is less than ~0.075 solar masses the density becomes so high that the gases become degenerate (at ~10 million grams per cubic centimeter) and further contraction is halted before the core can become hot enough to ignite bona fide hydrogen fusion through the proton-proton chain. The protostar then ends up as a brown dwarf.

Under the "cool" suboptimal temperatures found in the hearts of red dwarfs, the nuclear reactions that convert hydrogen into helium (the ppI chain) are not in equilibrium. This means that not all of the reactions happen with sufficient speed to produce helium-4 efficiently. Below 8 million K protons initially combine lazily to make deuterium, and then this burns much more readily[1] with additional protons to make helium-3. However, below this temperature the reaction that converts helium-3 into helium-4 occurs over such exceedingly long timescales that the helium-3 isotope builds up in the star. Only later on in the evolution of the red dwarf, when its core grows hotter, does helium-3 then complete the fusion pathway to produce helium-4.

Two other pp-chains, ppII and ppIII, are found in the Sun. These only occur when the core temperature exceeds 15 million K. For red dwarfs, with masses less than 0.2 M solar masses, this temperature is never reached at any stage in their evolution and these pp-chains are not utilized. More massive red dwarfs may burn hydrogen through these chains late on in their main sequence and post-main sequence lives as the abundance of hydrogen in their cores falls away.

Spectacularly Faint Dwarfs

Despite their generic faintness, some red dwarfs – including Proxima Centauri – belong to a class of variable stars called UV Ceti stars. Strong magnetic fields and a fast rotation stir and heat plasma in their star's atmosphere, producing powerful flares. These flares brighten the star by 1–3 magnitudes and generate powerful

---

[1] Proton capture on deuterium requires the least energy and this explains why it occurs readily in low mass objects (>13 Jupiter-masses).

X-ray and radio bursts. Indeed, the X-ray and radio flares produced by these stars are over 100 times as energetic as those on the Sun. It seems that these mice can occasionally roar. The precise mechanism by which these magnetic fields arise in the smallest stars is still a little unclear.

Interestingly, it isn't just the red dwarfs that sustain strong magnetic fields. We already know that giant planets such as Jupiter do so, and that these fields generate a lot of radio noise. However, it wasn't clear whether brown dwarf stars would be strong sources of radio emission as well – and some researchers had believed that they would be radio quiet. In the year 2000, *Nature* published an article laying this idea to rest. A young brown dwarf was found emitting bursts of radiation in the X-ray and radio wave bands. These emissions could only come from a hot corona of gas trapped and manipulated by the object's strong magnetic field. So even though this object had a cool and cloudy atmosphere, the internal magnetic field was still powerful enough to whip up a corona of hot gases above the cloud-tops. Small does not always mean placid.

It was assumed that stars as meek as the smallest red dwarf (with correspondingly weaker gravitational fields and lower surface temperatures) couldn't produce such displays. Certainly, it was thought that brown dwarfs would be the last place you'd expect to find high levels of activity. However, evidence suggests that even the smallest stars and their failed brown dwarf cousins can also generate a show as they condense out of the darkness. Size, it seems, is not everything. Consequently, many of the criteria used to segregate the brown dwarfs from their more massive peers seems no longer to apply.

A Primer of Stellar Evolution

Stars become progressively dimmer and cooler as we descend the main sequence from the position of the Sun. Stars like the Sun, occupying class G, make up a paltry 7 % or so of those in the universe, while cooler class K orange dwarf stars comprise perhaps 14 % of the universe's population of stars. The coolest stars in class M make up the vast majority of the remaining groupings, comprising at least 75 % of the universal stellar population.

Red dwarf stars are the universe's future. They are also its most abundant stellar factories. Their lives are measured in hundreds of billions or trillions of years – far in excess of the current age of the universe. Red dwarf stars are, therefore, the most important stars in the universe as potential abodes for life – or at least the sorts of life that we might most easily imagine.

Given the strong dependence of lifespan on stellar mass it is the M-class dwarfs that will reign in the universe's future. Like a scene from a socialist's starry-eyed dream, the reds will eventually inherit the universe in a 100 billion years' time. At this point the last K-dwarf star will have left the main sequence and evolved into a white dwarf and the blues will be but a distant, faded memory. That is, if nothing untoward happens to the present accelerating expansion of the universe. And although we can share some uncertainty about this, assuming nothing else happens, the current stellar population will evolve before eventually twinkling away into the blackness tens of trillions of years hence.

Red dwarfs are a bit of a research backwater – at least they were. Perhaps this was a consequence of their dismal luminosities, their diffuse and speculative lives, and the implication that "small-means-boring." Whatever the true reason, it meant that red dwarfs languished in the backwaters of stellar astrophysics for decades. At a rate comparable to the nuclear reactions powering their parsimonious lives, articles emerged steadily but at a geological pace over the years, with roughly two or three per decade from 1960 until the late 1990s. During the 1960s several researchers – notably Icko Iben and Peter Bodenheimer – had an interest in examining the future evolution of the lowest mass stars, but at the time, the physical properties of the stars were simply not well enough understood. By the 1980s, the physics had been sufficiently developed, but oddly, nobody returned to look at the future of the red dwarfs.

However, research into the evolution of red dwarf stars was resuscitated in 1996 with a series of research papers by Greg Laughlin, Fred Adams and Peter Bodenheimer. Astrophysicists, Laughlin, Adams and Bodenheimer saw this gap in the market as a fortunate opportunity to revisit the excitement of the 1960s when stellar evolution codes were first starting to really unveil the later life stages of the stars. Laughlin, Adams and Bodenheimer published a string of articles examining the fate of nature's least substantial stars. Consequently, this chapter is really dedicated to their efforts, without which subsequent research into the habitability of crimson worlds would be without foundation.

In parallel, researchers Gilles Chabrier and Isabelle Baraffe began work on the physical properties of red dwarfs, determining how their chemistry, luminosity and size varied with mass. Baraffe had previously worked on the structure and evolution of considerably more massive stars, so this was, at the time, a considerable departure. However, their efforts were fortuitous as these further advanced considerations of the habitability of red dwarf planets, none of which were known at the time. With the evolutionary models of Laughlin and colleagues, and the observations of the physical and chemical properties of these stars, subsequent work on habitability could focus on how these peculiar little objects could support living worlds. How would the relatively miniscule energy liberated by nuclear fusion impact on neighboring planets? The dim, unconventional fates of nature's least massive stars and their unique chemistry mean that red dwarfs have an influence on neighboring planets that is as unique as their lives. Of these factors, the dominant emission of infrared radiation, and a lifetime orders of magnitudes greater than the central engines of planets, mean that their contribution to habitability is unlike that of any other, more massive star.

Despite languishing in an astrophysical oxbow for most of the last few decades, red dwarfs *did* make a comeback. However, what remains peculiar is that no one appears to have published anything on the evolution of the orange, K-class, stars since 1993. Even here, this was an afterthought in a paper that concentrated on more massive stars. Thus, to write about the evolution of these stars much inference had to be drawn from what was known and published about the lower and higher mass stars. The description of the evolution of these stars was also inferred from observations of a special class of stars known as Horizontal-Branch giants that are found in star clusters, particularly the old, Population II globular clusters. It's a real shame, and rather perplexing, that in 2013 there is an entire class of stars for which no detailed evolutionary models exist. This is all the more true given the hunt for habitable worlds in orbit among what are 14 % of the total number of stars in the universe.

The Life of an "Ordinary" Red Dwarf

A red dwarf's uncharismatic lifestyle is its strength. With only microscopic exceptions, all complex life depends on the radiant energy from stars. However, in a little over 100 billion years (depending on the universe's pace of star formation), the only main sequence stars left will be red dwarfs with less than half the Sun's mass. It will be to these dim denizens of the galaxy any future human descendant will be forced to turn to for sustenance.

The life span of a star relates inversely and disproportionately to its mass. Weighing in between 0.075 and 0.55 solar masses (~75–550 Jupiter masses), the length of a red dwarf's life is measured more frequently in trillions, rather than millions or even billions of years. While the largest (0.55 solar mass) red dwarf will last a little over 100 billion years, the meekest red dwarf stars will last 12 trillion years. This figure compares very favorably with the Sun's paltry 10 billion years, or the inconsequential 3 million year long brazen reign of a 100 solar mass behemoth.

Similarly, the time it takes a red dwarf to reach the main sequence is extended. Although a proto-red dwarf will begin glowing within a few million years of commencing its collapse, this energy is derived for the most part from the conversion of gravitational potential energy to heat. Early on, during the first few tens of millions of years, additional energy is derived from the fusion of first deuterium to helium-3, then lithium-7 to helium-4, through proton capture. These are meager reserves, but they do allow the proto-dwarf to glow with a light comparable to the early Sun for millions of years. Only some time later does the core become hot enough to fuse the majority of the star's hydrogen to helium. Collapse then continues until the energy released by nuclear fusion matches that demanded by the inward gravitational pull of the star itself.

The time taken to perform the transformation from nebula to 0.1 solar mass star is in itself remarkable – 2 billion years, a time equivalent to nearly 1,000 generations of the universe's most massive stars. By contrast the Sun took roughly 10 million years to form itself from a collapsing and fragmenting gas cloud. In the time it takes a small red dwarf to stabilize its collapse through energy generated by the pp-chain, the Sun had nourished Earth for long enough to generate sophisticated biochemistry in millions of different types of bacteria. This in turn led to the formation of our breathable, oxygen-rich atmosphere. Were the Sun a red dwarf, any life dependent on it might barely have gotten going by this point, some 4.6 billion years later, and there certainly wouldn't have been anyone around to write this text or research its contents (Chaps. 9 and 10).

With an interior that convects from core to photosphere, red dwarfs make for efficient furnaces. For stars with masses less than a quarter that of the Sun only 1 % of the star's mass will be hydrogen at the end of its life. The less efficient Sun will have used less than 60 % of its total hydrogen fuel reserve by the time it dies – the unspent fuel largely expelled as a beautiful planetary nebula.

The story of a red dwarf is one of efficiency coupled to an intense proclivity for parsimony. Nearly all the fuel is exploitable; there is absolutely no rush to get through it, and the ultimate product, helium-4, can wait an eternity for its synthesis. A 0.1 solar mass red dwarf, like Proxima Centauri, will take its time to synthesize helium-3, and then nearly forever to convert this into helium-4. There is no rush. It is the thriftiness of red dwarfs that make them such ideal abodes for developing life. Once life has a foothold on a neighboring world, opportunity drifts onwards like a feather, lazily spinning upon the breath of time.

At birth, a star with one tenth the mass of the Sun has a surface temperature of around 2,200 K – broadly equivalent to burning thermite. Its core hovers around 2.8 million K, considerably less than that of the Sun. Compared to the Sun, the 0.1 solar mass red dwarf puts out a paltry 1/3,000th that of our star, meaning that to our eyes such a star would be invisible unless it was located within 1.3 light years of Earth.

The evolution of the red dwarf is driven by a delicate stepwise synthesis of helium-4. Within a cool red dwarf these reactions run out of kilter, and the final step that manufactures helium-4 from helium-3 lags far behind the others. That is not to say it isn't happening. It's just very slow in comparison. The effect of this is to allow helium-3 to build up in the red dwarf through a sizable chunk of its life – something that won't happen in the Sun.

As the 0.1 solar mass red dwarf traverses the first 1.4 trillion years of its life helium-3 builds up until it occupies almost 10 %

of the mass of the star. Throughout this period the little red dwarf slowly expands and heats up slightly. It does this because the core is steadily growing hotter as the amount of helium-3 increases. With more helium-3 around, the reactions that make helium-4 can go faster and the star can generate more energy. For any planet in the star's habitable zone, the effect is slight. The star would appear slightly brighter and bluer, but the temperature at the surface of the planet wouldn't change much.

In somewhat more massive red dwarf stars (0.12–0.16 solar masses) the core temperature actually falls during the equivalent phase. As the amount of helium-3 increases, the upsurge in energy production is enough to expand and cool the core somewhat. However, once this phase ends, the amount of helium-3 must fall, and the star enters the second performance of its show. As the trillions of years advance the amount of helium-4 steadily increases, and the core temperature rises to over 5 million K. At 4.4 trillion years of age helium is now more abundant than hydrogen, and the final struggle to maintain stability gets under way. The red dwarf's surface will now have heated to 2,500 K, and the star will have brightened to approximately 1,000th that of the current Sun. It's still a red dwarf, but it is slowly becoming blue.

Throughout its main sequence life convection keeps the composition of the star homogeneous, steadily removing hydrogen from the outermost parts of the star, dragging it into the interior, while freshly minted helium is depositing away from the core in the envelope. Once helium-4 dominates the mass of the dwarf, the amount of energy that can be generated by fusion reactions decreases. The red dwarf is then compelled to shrink its interior and raise its temperature further. This increases the rate of nuclear reactions and allows the red dwarf to, at least temporarily, balance its books once more.

As the core collapses, the helium slowly flooding the envelope of the star makes it increasingly transparent to the radiation generated in its core. This effect is twofold. As hydrogen is turned into helium there is less of this element to soak up radiation from the core. Helium is far less effective at absorbing radiation in the stellar envelope than hydrogen. Moreover, as hydrogen is consumed there is less of it to form organic molecules such as CH – the simplest compound of carbon and hydrogen. As these

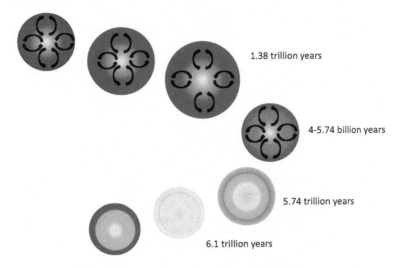

1.38 trillion years

4-5.74 billion years

5.74 trillion years

6.1 trillion years

FIG. 3.4 Changes to the internal structure of a red dwarf star as it ages

compounds are also effective absorbers of radiation, their loss means that the envelope, as a whole, cannot absorb radiation as effectively as before. Thus, more radiation simply leaks out of the star without depositing its energy in the envelope, and the star must, therefore, contract. Contraction means it becomes hotter and a hotter star has less organic molecules to grab hold of the radiation and less hydrogen gas in its neutral form (Fig. 3.4).

With an increase in the rate of nuclear reactions and a lower opacity, the dwarf brightens further and grows steadily hotter, even as it shrinks in diameter. After 5.74 trillion years, the star has a surface temperature of 3,450 K – approaching that of the most massive M-class dwarf stars of today – and a luminosity of approximately 0.5 %, or 1/200th, of the present Sun. At this point only 16 % of the star's mass is hydrogen and the end is finally approaching.

After 5.74 trillion years helium dominates the composition of the star. The stellar material is now so transparent to radiation that convection is no longer the best means of transporting energy through the star. A direct flow of radiation is the more effective means of energy transport. With energy now able to escape directly, rather abruptly star-wide convection ceases. The one unique feature of the red dwarf – its whole star convection – is now over, and the structure comes to resemble that of an old Sun. The red dwarf now has a helium-rich core overlain by a thin layer of hydrogen.

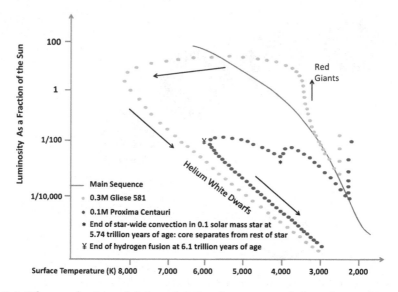

FIG. 3.5 The evolution of 0.1 and 0.3 solar mass red dwarf stars based on the work of Greg Laughlin, Peter Bodenheimer and Fred Adams. The 0.1 solar mass red dwarf (Proxima Centauri) never becomes a red giant, while the more massive Gliese 581 dwarf does. Both red dwarfs ultimately fade away as helium-rich white dwarf stars

Although convection has ended in the core, it continues in the thin hydrogen-rich shell surrounding it. This stretches out to the star's surface. All the hydrogen remaining in the now stagnant core is rapidly burned to helium. With this sudden change in structure, the star abruptly contracts a few percent and dims. This is the end of the red dwarf's main sequence and is evident in Fig. 3.5.

After the core runs out of fuel, hydrogen will continue to burn in a shell surrounding it, which is, in turn, overlain by a convective outer hydrogen-rich shell that retains the overall abundance of hydrogen that remained after whole-star convection ended. The helium core then simply contracts and grows hotter as it gains mass from the hydrogen-burning shell, much as it will do in the Sun in 5 billion years. The hydrogen-burning shell source will burn outwards as the helium core contracts. This progressively consumes the remaining hydrogen until 96 % of the former red dwarf has been transformed into helium-4. During this final protracted phase its envelope contracts and heats rapidly (Fig. 3.6). The Sun will expand during the equivalent phase, but the old red dwarf will not.

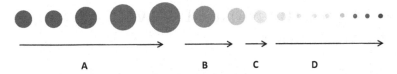

FIG. **3.6** The evolution of a 0.1 solar mass red dwarf. During phase *A* the star steadily coverts its hydrogen to helium, and the increase in energy efficiency causes the star to slowly expand and heat up. During phase *B* the amount of helium in the star reduces energy efficiency, and the star is forced to contract once more and heat further. After 5.7 trillion years the star's core stops convecting, and the remaining fuel is burned rapidly. Finally, after 6 trillion years most of the hydrogen has been consumed, and what remains of the star shrinks to become a white dwarf. Any remaining hydrogen in the star's outer layers is used up before the star begins to cool and fade into invisibility

Eventually, some 402 billion years after the core ceases convecting, the star will have heated to roughly 5,800 K, making it somewhat hotter than that of the present day Sun. Although its surface is now hotter than our star, it will only be a little over 100th as bright. This is as good as it will get for the 0.1 solar-mass red dwarf. The core will have a temperature of 12 million Kelvin (again, comparable with that of the present Sun) and, more importantly, will have become fully degenerate. In this state electrons are confined to particular energy levels, and they take up the strain against further compression. Energy transfer by radiation gives way to conduction, much as it would in a solid material such as a metal. In the degenerate state, electrons in the dense helium plasma resist further compression, and any further contraction that does occur does not increase the temperature. Consequently, although the envelope can contract further, the core is as dense and as hot as it can get.

The antiquated (former) red dwarf now "turns the corner" on the Hertzsprung-Russell diagram. The dwarf will spend the next 134 billion years contracting, cooling and dimming into invisibility. As it cools and fades from view, the star's hydrogen burning shell will gradually wink out. A colossal 6 trillion 279 billion years after the dwarf reached the main sequence (and eons after the Sun's shriveled remains became as cold and desolate as the space surrounding them), the spent helium dwarf will fade from view forever.

The Fate of the Most Massive Red Dwarf Stars

What happens to red dwarfs more massive than the example considered here? For example Gliese 581 and Gliese 667C have masses three times greater than the example explored above. Will their lives differ in some dramatic fashion from the smaller red dwarf stars? To précis their fate, in general, as the mass of the star increases, these more massive red dwarfs become increasingly like the stars we see evolving and dying today. Most significantly is the appearance of a red giant phase.

More massive red dwarfs, including the 0.3 solar mass stars Gliese 581 and 667C, follow broadly similar evolutionary paths as their smaller cousins. However, the main differences are that the stellar core ceases convection at progressively earlier times in the star's history as the stellar mass is ratcheted up. This is important, as it means that proportionately less of the star's hydrogen is used up, with the leftovers locked into the envelope. Furthermore, as the mass increases, the dwarf stars become proportionately hotter once the core ceases convection. The most massive of the aging red dwarfs will turn distinctly blue. A typical 0.2 solar mass dwarf will have a surface temperature above 10,000 K (hotter than Sirius) when it, too, "turns the corner" on the Hertzsprung-Russell diagram.

The calculations of Laughlin and colleagues showed that red dwarfs weighing in excess of ~0.16 solar masses underwent expansion once the radiative cores appeared. It seemed that these stars were the lowest mass red giants that the universe will produce. In the case of a 0.20 solar mass star, the expansion of the envelope is substantial. The star expands to more than ten times its original diameter. Although such expansion would only endow these little stars with a girth comparable to the present Sun, it suggested that the smallest red giants will be formed from stars with masses greater than approximately a fifth that of the Sun. Stars less massive than this limit will grow neither red nor giant. Less massive stars would simply heat up and brighten, becoming bluer, before dimming and finally dying.

Why do these more massive stars become giants rather than simply get hotter? The key lies in the amount of hydrogen and helium in their envelopes. The more massive red dwarfs, those that retain more hydrogen in their envelopes, are able to trap more of the larger amount of energy pouring out of the aging stellar cores. A relatively lightweight 0.16 solar mass star increases its luminosity 140-fold with little expansion. The bulk of the increased output from the core is accommodated by more vigorous convection, a higher surface temperature and only a very modest expansion.

Conversely, a star with 0.25 solar mass cannot dump the extra energy from its core late on in its life. It simply creates too much. Instead, as the star increases its luminosity a 1,000-fold, eventually becoming brighter than the present Sun, the envelope is forced to expand. The additional energy can't simply be removed by heating up the star's outer layers. The star has to expand and increase its surface area to shed the extra energy from its core. This is the only way the star can keep its interior in equilibrium.

The driving force behind this important change in evolutionary path appears to be the temperature of the degenerate core. In red dwarfs less than ~0.2 solar masses the degenerate core maintains a constant temperature even as the hydrogen shell source burns outwards and adds mass. The ephemeral neutrino is the source of this consistency. As degeneracy sets in particles called neutrinos carry energy out of the dense core, thereby allowing it to cool effectively. This means that these small stars never generate so much oomph that their envelopes are overloaded with energy. However, stars with more than 0.2 solar masses develop progressively hotter, degenerate cores as their mass is increased through shell burning. In these more massive red dwarfs, the core mass goes up as the surrounding hydrogen burning shell adds extra helium ash. This bumps the temperature up, which in turns up the rate of nuclear reactions. However, neutrinos can't dump the extra energy so the star has to expand to do it instead.

Thus it is the hotter and more luminous core that drives the development of giant-hood. Moreover, the smallest red dwarfs flood their envelopes with helium as convection carries it steadily from the hot core. The presence of abundant helium lowers the ability of the envelope to absorb energy, allowing it to escape more readily. Thus a hotter core and a less transparent envelope mean stars with masses greater than one fifth that of the Sun become red giants, while those less substantial simply get hotter and become white dwarfs directly.

Red giants end up in a spiral of expansion because of one final reason. As the temperature falls with increasing size, molecules and grains of matter begin to clog up the stellar surface. This only gets worse as the temperature lowers. Overall the envelope of a star becomes less opaque to radiation as the temperature drops from 5,000 K to 2,000 K before increasing rapidly once more. Therefore as the temperature initially falls more energy is able to leave the surface of the star by radiation. Above 5,000 K H^+ and H^- contribute to how opaque the material is, while below 2,000 K metals and complex organic compounds become significant blankets for radiation.

As the red dwarf evolves and its luminosity increases, the star tries to remain in the region of lowest opacity, where energy can most readily escape its surface. However, for the more massive red dwarfs, the temperature of the envelope increases towards 5,000 K and the opacity is eventually forced upwards. This reduces the capacity of the stellar surface to dump energy into space. Since less energy is able to escape, the star has to expand to remain in balance. Hence, both the increase in luminosity and the decrease in the temperature of the photosphere force giant-hood upon the more massive dwarfs.

The Missing Piece

One piece of evidence still lacking from currently published work is the extent to which red dwarf stars lose mass through stellar winds. The Sun loses one trillionth of a solar mass per annum. This is nowhere near enough to affect its fate during its main sequence life. However, is this rate of mass loss applicable to red dwarfs? If it is too high, a red dwarf might be reduced to a flickering ember by loss of mass in its expansive lifetime. At present the true rate at which red dwarfs shed mass isn't known. However, as long as it is less than ten trillionths of a solar mass per year there will be little if any effect on their future lives and that of any planets that are nurtured by them. Remember if a red dwarf loses mass at the rate the Sun does it would be reduced to a wispish ghost in one trillion years. Nuclear reactions would end long before this. If it is ten trillionths of a solar mass per year no red dwarf will live past a trillion years. The loss of mass will extinguish its central furnace long before the fuel could be consumed. The rate isn't known at all, beyond simply stating that it is low. Unfortunately, without this the total habitable lifespan of a planet orbiting the star can't be known either.

Although the research of Greg Laughlin and co-workers didn't take into account mass loss, it can be expected that once the larger red dwarf stars evolve onto the red giant branch (RGB) they will experience greater mass loss, just like their larger and currently faster evolving and more massive cousins. Consequently, subsequent stages in their evolution will probably be truncated. The models that Greg Laughlin followed ran into severe difficulties when he tried to follow the subsequent evolution of the dwarf as a red giant. Therefore, not much can be said at present about what would happen next. Further enlightenment must await detailed modeling of stars in the 0.3–0.5 solar mass range.

To sum up, at masses greater than a fifth that of the Sun the behavior of the red dwarfs begins to resemble that of more massive stars that we are used to seeing evolving today. At the close of the main sequence a 0.25 solar mass star leaves 50 % of its hydrogen fuel unused (compared with 16 % for the 0.1 solar mass star described above). Although these red dwarfs have a long way to go before they carry out these transformations, even though they will never be able to perform the internal alchemy that characterizes their more massive brethren, they will at least give the outward impression that they are trying to. The red giants of the future may be smaller than those of today's universe, but they will continue to grace the skies for more than a trillion years to come.

Flings for Mid-Range M-Dwarf Stars

Calculations carried out in the 1970s by Ronald Webbink suggest that stars with bulk of around 0.2 solar masses may have a few tricks to play before they ultimately fade as white dwarfs. These stars will become red giants, but as their hydrogen envelopes blow away in the stellar breeze, the remaining hydrogen layer will shrink in mass, contract and become degenerate. Hydrogen fusion in these hot shells is through the carbon-nitrogen-oxygen (or CNO) cycle, strongly dependent on temperature. As the remaining hydrogen layer contracts, its temperature strongly increases. This has the potential to lead to explosive burning. Webbink suggested that these ancient red dwarfs of the future may go through a final, explosive phase where the remaining hydrogen layer convulses, ejecting much of itself into space in nova-like outbursts. More massive red dwarfs avoid this by maintaining stable hydrogen burning until their remaining reserves are blown away, presumably as diminutive planetary nebulae.

Whatever the ultimate truth, these stars are doomed to a slow fade out. Their hydrogen exhausted, their core will begin to cool, and the remnant will complete its life over tens of billions of years as a slowly fading helium white dwarf.

The Evolution of Orange K-Dwarfs

Spanning a narrow range of masses, between the parsimonious red dwarfs and the more recognizable home comforts provided by the yellow, G-dwarfs lie the orange K-dwarfs. Their masses vary from roughly 0.6 solar masses to 0.75 solar masses, their surface temperatures from 3,500 to 4,900 K and it is estimated that they comprise about 14 % of the total number of main sequence stars in the galaxy.

Like their more massive brethren, these stars will be able to burn helium when they eventually leave the main sequence. And like the red dwarfs, the universe is still too young for any orange K-dwarf to have left the main sequence and begun its journey to the grave. At present the least massive stars evolving are metal-poor 0.8 solar mass stars in the galactic halo. These are yellow G8V dwarfs. It will be approximately another 5 billion years before the most massive galactic K-dwarf leaves the main sequence. Therefore, like the red dwarfs before them, this chapter's evolutionary models are entirely theoretical and won't be confirmed or refuted until the Sun has swollen to become a red giant and every library on Earth has been thoroughly incinerated.

What we do know is that in terms of their future direction, the K-dwarfs will follow the same general route taken by more massive objects such as the Sun. An orange K-dwarf will remain on the main sequence for anywhere between 15 and 100 billion years, with the most massive orange dwarfs leading the shortest lives. During this time the majority of energy will be generated by the pp-chains that generate output in red dwarf stars. The most massive K-dwarfs also produce a miniscule ($\ll 0.1$ %) additional amount by the CNO cycle. However, given its trivial role it can be neglected.

A K-dwarf will take between 50 and 300 million years to contract onto the main sequence. Aside from the shorter period of time taken, the K-dwarf will follow the same series of pre-main sequence events as the smaller M-class dwarfs.

Protostars ranging in mass from ~0.3 solar masses up to ~2 solar masses go through a stage called the T-Tauri phase. Here the contracting protostar is racked by violent eruptions, caused by a mixture of debris impacting on the still contracting star, and by flaring associated with the star's strong magnetic field. Magnetism is initially strong because the young star rotates rapidly and experiences strong convection within its bulk. As the star ages the rate of rotation slows and the magnetic field declines. However, work by John Stauffer (Caltech) suggests that K-dwarf stars in the relatively youthful Pleiades cluster may have up to half of their surface covered in star spots.

In the late 1990s and subsequent decade, observations had shown that the Pleiades K-dwarfs were too blue for their masses, but also about half a magnitude to faint to fit onto the main sequence. This spectral discrepancy could not be explained by below average metallicity. When compared with the older Praesepe cluster, the Pleiades stars were oddly dim; something was clearly amiss. Bizarrely, at longer wavelengths the Praesepe and Pleiades matched one another. This implied that more than 50 % of the surface on the Pleiades K-dwarfs was covered in cool, dark star spots, with those in the older cluster less afflicted. The apparent conflict of color and brightness was thus resolved because a larger area was dark for most of the time but periodically generated flares that produced bursts of ultraviolet and X-ray emission. Indeed, the suggestion from additional research was that all young K-dwarfs exhibit severe acne as a consequence of strong, localized surface magnetism.

The T-Tauri stage is also associated with the production of outflows or jets in some stars. As material swirls through the accretion disc and crashes onto the surface of the developing star, interstellar magnetic field lines get drawn inwards and twisted. Inward-flowing material carries with it a significant amount of angular momentum, effectively a measure of spin. The protostar may shed this excess momentum by flinging material outwards from the accretion disc along the stellar rotation axis, probably with the aid of a strong magnetic field. A few percent of a stellar mass may be expelled in this way, generating the characteristic jets and glowing Herbig-Haro objects. These odd, luminous apparitions are generated where the outward moving jets impinge on the surrounding nebula. Indeed one model suggested that the infant Sun may have shed as much as 0.07 solar masses of material in this way.

Moreover, in addition to mass loss through proto-stellar jets, strong stellar winds also characterize these immature stars, and these, too, can remove additional angular momentum, slowing the overall speed of stellar rotation. Both of these effects tend to diminish early on before the star has settled onto the main sequence. Once the star has stabilized on the main sequence, any jets will have subsided and mass loss through stellar winds will decline to one trillionth of a solar mass per year. This is a very comfortable level that wouldn't affect the further evolution of the star. The rate of mass loss is then far less than the rate of nuclear reactions in its core.

Magnetic fields are generated within K-dwarfs in the same manner as the Sun – the so-called $\alpha\omega$ (alpha-omega) effect. In essence this complex mechanism relies on two factors. The first of these is determined by the structure of the star. The central regions are radiative and rotate rapidly as a single entity relative to the outer convective shell. Within this layer the vertical motion of gas, through convection, is modified by the rotation of the star, a process called the Coriolis effect, which is discussed in Chap. 6. Within stars such as the Sun, these conflicting forces cause the fast-rotating radiative core to slide against the base of the more sluggishly rotating outer convection zone. This generates the complex zone called the tachocline, which was discussed earlier.

At stellar birth the convecting outer layer of the star retains and concentrates the interstellar magnetic field. Through a self-reinforcing mechanism that is still poorly understood the magnetic field is propagated outwards through convection to the stellar exterior and manifests itself in localized cells. These generate star spots and other magnetic phenomena. Magnetic fields generated beneath the photosphere heat the outer atmosphere, generating a hot (1–2 million K) corona that is detectable through X-ray emission.

Since star spots may reduce the amount of visible light emitted by the star during its more formative years, this could result in large swings in temperature on any planet orbiting it. This is especially true of red dwarf planets, where the amount of incident visible radiation is already low. However, a planet orbiting within the habitable zone of a K-star should expect strong variation in temperature early on if the number of star spots is high, as the overall amount of radiation emitted by the star's surface will be greater than that emitted by a smaller red dwarf.

We have also seen that star spots are sources of violent flaring activity. Flares release large quantities of X-rays and ultraviolet light. Once more, any planet impinged upon by these outbursts will need the protection of a strong magnetic field if its atmosphere and any life are to be shielded. However, unlike the tidally locked worlds of red dwarfs, the vast majority of planets orbiting within the habitable zone of an orange dwarf should rotate at least slowly relative to its stellar host. The stellar habitable zone extends outside the tidal-lock zone for these more massive and more luminous stars. Stronger rotation has two important effects that are discussed more fully later on: the presence of stronger atmospheric circulation driven by the Coriolis effect (Chap. 6) and the greater likelihood of a strong planetary magnetic field. Both rely in part on the rate of rotation of the planet and its iron core. These caveats aside, star spots decline in frequency with age as the rate of stellar rotation slows and magnetic field declines. The more mature red and orange dwarf citizens of the galaxy are expected to suffer less in the way of these bipolar outbursts and hence pose less risk to their progeny.

The K-dwarfs are very similar in many regards to the more massive but rarer G-dwarfs such as our Sun. It is only their intrinsic faintness that casts them apart. At ~0.6 solar masses the least massive K-dwarf (or most massive M-dwarf) is only 1 % as bright as the Sun. At their most massive, the K-dwarfs are approximately a third as bright as the present Sun while on the main sequence.

However, it is worth pointing out that this luminosity is only half that of the newborn Sun and that one of these stars wouldn't look terribly out of place in our solar system.

After several tens to a few tens to hundreds of millions of years the K-dwarf settles onto the main sequence at its zero-age main-sequence, or ZAMS. After this point the star remains stable for the next few tens of billions of years. During this period, the pp-chain converts hydrogen to helium in a small radiative core that varies in size from 1/5 to 1/4 the total stellar diameter (Fig. 3.3). The smallest K-dwarfs (a little over 0.55 solar masses) have convection in roughly the outer two-thirds of their structure. The thickness of the convective shell decreases quite rapidly with increasing mass. A K0 star, with approximately 0.75–0.8 solar masses, has convection confined to a little less than the outer third of its mass. As with the most massive red dwarf stars, this mode of construction means that the core is isolated from the rest of the star. Like the smaller red dwarfs with masses above one quarter that of the Sun, this means that the fuel in the envelope makes little contribution to the life of the star throughout the bulk of the star's life. Thus when the abundance of hydrogen in the core falls below 0.1 % the core is unable to generate enough energy to support itself, and promptly begins a phase of rapid contraction under the influence of gravity.

While on the main sequence a K-dwarf, like its smaller red dwarf cousins, gradually changes its behavior. As the core progressively uses up its supply of hydrogen, the efficiency of hydrogen fusion per gram of stellar mass is lowered. The star compensates by drawing on its immense supply of gravitational potential energy. The stellar core slowly contracts, heating and raising the rate of hydrogen fusion. Indeed, like the M-dwarfs, all stars pull this same trick to compensate for gradually dwindling energy resources. The increase in the rate of energy production in turn causes the star to gradually brighten and become hotter in the same way as the lowest mass M-dwarfs we looked at in the previous chapter. On the HR diagram the star moves upwards to the left, along the main sequence, brightening and becoming steadily bluer with time. This upward trend is terminated when the core finally exhausts the available hydrogen. For the M-dwarfs, this process can take several trillion years and results in the M-dwarf eventually becoming a yellow G-dwarf – albeit much smaller than the ones we are used to today. Similarly a typical K-dwarf will eventually end up as hot as the present-day Sun prior to leaving the main sequence.

Once the core exhausts hydrogen, energy output from the core briefly subsides and the star is forced to contract and heat up. The star makes a sharp turn to the left on the HR diagram. This temporary phase ends when the heat released by the contracting core causes the hydrogen lying immediately above it to ignite in a thick shell.

Re-invigorated by this new supply of nuclear energy, the overall stellar energy flow increases while the core contracts, gains mass and grows hotter. Once the shell source is ignited, the star begins expanding in response to the extra energy available, just as it did in the red dwarf. During this phase of core contraction, the volume of the core shrinks from a diameter approximately equivalent to that of Jupiter down to an object approaching the size of Earth (roughly a 100-fold contraction) while its density rises to ~60 million grams per cubic centimeter.

During its main sequence life, hydrogen fusion was generated exclusively by the sluggish proton-proton chain. The rate of this reaction has a fairly weak dependence on temperature and so will only slowly increase as temperatures go up. However, as the star slowly expands into a sub-giant and then a red giant, hydrogen fusion burns at a much greater rate in the shell through the carbon-nitrogen-oxygen (or CNO) reactions. These reactions have a very steep dependence on temperature, and so as the temperature continues to rise as the core contracts, the rate of nuclear reactions accelerates as well. This is positive feedback: higher core temperature lead to faster nuclear reactions. This, in turn, causes acceleration in the production of helium. More helium in the core makes it heavier, and it contracts and heats faster; and so on and so forth until something gives.

Over the course of several hundred million to a few billion years the small K-star takes on the appearance of a red giant. Its surface cools, molecules and grains begin to condense in its outer layers and strong stellar winds develop that begin to whittle mass away from the star.

Which Stars Burn Helium?

The orange K-dwarfs and the most massive M-class red dwarf stars are in a precarious position on the HR diagram. Stars lose mass through stellar winds, and this rate accelerates greatly as they expand and become red giants. If the mass loss takes them down to much less than 0.45 solar masses, their helium core will not become hot and dense enough to fire up. K-class stars begin life with a low mass, which means that if they lose enough they could end up with too little to ignite their helium store as red giants. These failed giants would then fade away as helium white dwarf stars, much like their smaller red dwarf cousins.

The helium-burning limit is theoretical, but we do know of quite a few, select stars with masses around this limit (0.46–0.5 solar masses) that do burn helium. However, for the smallest K-dwarf stars and the largest red dwarf stars, the amount of mass lost during the red giant stage might prevent them from attaining a massive enough core of helium. For the largest single K-stars this risk is easily avoided, and they will produce a red giant star with a helium core of around 0.5 solar masses. However, at the low end of the range these small K-stars (and most massive red dwarfs) could encounter some difficulty. Therefore, red dwarfs with spectral classes M0 and M1, such as Barnard's star, may just grow their helium cores sufficiently to ignite them. Smaller red dwarfs will certainly fade away as helium-rich white dwarf stars. A K-dwarf may lose as much as one fifth of its mass (20 %) during its ascent of the red giant branch. For a star with 0.6 solar masses to begin with, this takes its mass down to 0.48 solar masses – just above the helium burning limit. Such stars may well ignite their helium fuel, but would they be red giants? Any further mass loss and the star must become a helium white dwarf.

For the more massive K-dwarfs – those with masses in excess of 0.6–0.65 times that of the Sun – the star ascends the red giant branch. When its luminosity is roughly 2,000 times that of the present Sun, its core will hold approximately half a Sun's worth of helium gas in a strongly degenerate state. The star will have expanded a 100 times its original diameter, and the surface will have cooled to between 2,500 and 3,500 K. Rather abruptly helium will then ignite. The degenerate helium core will be resistant to change. Heat will then build up from the nuclear reactions, causing the temperature to rise and the rate of reactions to accelerate. This causes a violent eruption within the core, and the stellar output rises to 100,000 times that of the present Sun. The core shudders and violently expands, cooling until the pace of nuclear reactions subsides and normality returns. The consequences for the star on the outside are unclear. The star may expand and even shed some of its mass, but soon thereafter, the star will settle down once more to a new state. What is this like?

A helium-burning giant is smaller, hotter and denser than its red giant branch progenitor. These stars are orange or yellow in color and have diameters perhaps 30 times that of the Sun. Their luminosity is perhaps 100 times that of the current Sun, placing them on a strip of stars extending across the middle of the HR diagram. At present this horizontal branch is populated by low mass, metal-poor stars. In the future the lowest mass helium-burning stars will reside here, eking away at the last of their fuel.

Stars achieve this hotter, reinvigorated state by shrinking their hydrogen-rich envelopes in response to the expanding and cooling core. This is a result of both the conservation of angular momentum and a reduction in the amount of energy generated by the core. Why so? As the core expands outwards angular momentum must be conserved, and so the envelope shrinks in return. Moreover, as the core expands it cools down. Not only does this mean a reduction in the rate of helium fusion but a cooling of the hydrogen-burning shell to the point that it effectively shuts off altogether. Since hydrogen fusion generates roughly ten times the amount of energy per gram of fuel as helium fusion, the overall energy output from the core falls dramatically once helium fusion commences. In turn this means that the star has to do less to shed the energy generated. The envelope can contract and energy can be radiated from a smaller surface area than needed in the preceding red giant phase.

Helium fusion lasts roughly 150 million years and is fairly constant for all low mass stars, the Sun included. This is because all stars with overall masses less than 2.25 times that of the Sun ignite helium once their core grows to 0.44 solar masses. Since the mass of the helium core in all low mass stars – down to the helium-burning limit – is the same, the time they spend burning this fuel will also be the same. More massive stars have sufficiently bulky cores to begin helium fusion very soon after they leave the main sequence. The mass of their helium core varies with the overall mass of the star, and thus the rate of helium fusion depends on the mass of the helium core at this point. In turn more massive helium cores burn their fuel faster than small ones, with a time span ranging from 100 million years in stars with 2.25 times the mass of the Sun to less than a million years in the most massive stars.

The Second Ascension

Once helium fusion is over the stellar core is filled with roughly 0.5 solar masses of carbon and oxygen. Helium fusion winds down and becomes confined to a shell surrounding the inert carbon-oxygen core. The star may make a brief loop to higher temperatures before the inert core begins to collapse under its weight. With higher temperatures in the outer parts of the core, the outer hydrogen-burning shell re-ignites, and the energy output soars once more. The star begins a second and final ascent of the red giant branch, this time at slightly higher surface temperatures. This is the asymptotic giant branch.

As the star ascends, helium fuel is constantly added to the outer core of the red giant. At some point, the mass and density of this layer becomes sufficient to ignite the fuel. At this stage, the star enters its final phase of active evolution. Helium fusion expands and cools the helium shell and the hydrogen burning layer above it, until it switches off once more. A cycle then ensues with helium fusion exhausting the helium shell and an inert hydrogen-burning layer above. This phase alternates with phases when the helium is exhausted and the hydrogen burning shell is active and re-stocking the helium layer. Each breath of the hydrogen shell generates enough energy to re-expand the star, while phases of helium fusion are associated with shrinkage, higher temperatures and lower luminosities.

Each thermal pulse, as these alternating phases are known, lasts for several thousand years until the hydrogen layer is exhausted. Perhaps unexpectedly, fuel exhaustion does not come from its consumption.

As the star's core increases in mass so does the luminosity of the star, so it continues its ascent of the red giant branch. How far a star advances back up the red giant branch will depend upon how much hydrogen it retained in its first red giant phase. At present no one has modeled this, and the future evolution of these stars is unclear. What we can say is that throughout this time the strength of winds blowing from its surface increases. This isn't a steady process, in part because of thermal pulses that cause periodic changes in luminosity, but mainly because of instability in the envelope of the giant star.

As the temperature falls at the surface molecules begin to form, and grains of silicate or carbon compounds begin to form. These absorb radiation from the star effectively and accelerate, moving outwards away from the center of the star. These molecules and grains drag on the hydrogen and helium in the envelope, pulling them away from the star. The star, therefore, develops a very strong wind, carrying mass away.

At temperatures less than 6,000 K hydrogen begins to recombine, and this also converts thermal energy to kinetic energy. At some distance below the stellar surface recombination and ionization of hydrogen causes the amount of energy released by the star to fluctuate. When hydrogen forms from ions and free electrons energy is released, which radiates outwards. This process allows the envelope to relax and contract inwards. As the envelope contracts, the density of the gases rises, and ionization begins once more. This process cycles over periods of months to years, depending on the mass of the star. Each wave of expansion drives further mass loss.

In the end, the combination of ionization pulses and the dragging effect of molecules and grains strips the entire envelope from the star, leaving a hot, dense core topped by a thin shell of hydrogen. With the stellar fuel gone, the star has no options left and must leave the giant branch. As luminosity is driven by the mass of the core, the remnant star then heats up, moving left across the HR diagram towards higher temperatures while maintaining its overall luminosity. Once the temperature exceeds 10,000–15,000 K at the stellar surface ultraviolet light becomes the dominant type of radiation.

Ultraviolet light is an effective ionizer of gases, driving a final, beautiful swan song. The expanding shells, lit from below by the hot core begin to glow, and a planetary nebula begins to take shape. Over the period of a few 100 years the increasing ultraviolet light, coupled to the thinning, dispersing envelope causes abundant fluorescence. Chemical elements present in the dispersing gases glow red for hydrogen and blue or green for oxygen and nitrogen. Other elements forged in the stellar furnace cast off some of their electrons, and the process of recombination displays the elemental fingerprints of these, too.

For 30,000–50,000 years these gases continue to glow until the nebula is perhaps a light year wide. Continued expansion dilutes out the gases until their combined light is too faint to see and the planetary nebula fades quietly and slowly from view. What's left is a hot stellar core – a white dwarf. By the time 100,000 years has passed the remaining stellar fuel is exhausted, and the white dwarf cools and contracts slightly, its fate to slowly dwindle over tens of billions of years. Within the white dwarf electrons provide pressure, holding the remnant up against the inward pull of gravity. And so, the star remains as an Earth-wide, cooling ember well into the distant future of the universe. All K-dwarf stars will share this fate.

For a handful of such dwarf stars sufficient helium will have accumulated in these final days to ignite once more. For a brief period – brief by stellar standards – these white dwarfs will enjoy a transitory renaissance. The wave of nuclear fusion will inflate their remaining hydrogen and helium-rich layers, driving the star back towards the red giant branch. Vigorous fusion of both fuels generates some peculiar surface chemistry. These born-again giants have surface temperatures of several thousand degrees, glow yellow-white in color and show periodic dips in brightness as carbon soot condenses in the expanding gases. Sakurai's object did just this in 1996. More than 15 years on, its surface is now hidden by dense clouds of soot and other matter created and dredged out of its interior.

The fate of these stars is to return to the white dwarf tract once more, their bulk depleted in hydrogen, as hot blue objects. Once their helium fuel is gone, they begin to cool again, perhaps surrounded by an inner helium-rich planetary nebula, buried in turn within a faint, outer, and much older hydrogen rich shell. The time span for this final fling is probably measured in decades – a mere blink in the cosmic eye.

The Fate of the Smallest K-Dwarf Stars

For the lowest mass K-dwarf stars a race begins, the outcome of which determines their ultimate fate. The nature of their destiny depends on the amount of mass they can hang on to while they swell as red giants. Their initial path mirrors that of more massive K- and G-class stars. However, at some stage the amount of mass they are losing drives them down a path approaching the lower mass limit for helium fusion.

The smallest K-dwarf stars will ascend the red giant branch, following the fate of their showier and more massive cousins. However, as the core grows, there will be a substantial reduction in the mass of the hydrogen envelope as it is both consumed from below and lost from above. Eventually, the mass of the envelope is so low that radiation take over energy transfer from convection, just like in the decrepit red dwarfs we met earlier. As this happens the envelope of the giant will shrink back, heat up and the star will become progressively hotter. These hydrogen-deficient stars will lie to the left of the red giant branch in the region associated with helium burning (the horizontal branch). All the while they will be constantly migrating leftwards towards higher temperatures. As hydrogen is still being fused into helium, the mass of the core continues to grow. Normally the luminosity will rise as the mass of the helium-rich core goes up. However, these stars are so deficient in fuel that the rate of hydrogen fusion stays low, and any increase in the rate of burning is offset by the progressive contraction of the star. Consequently, these stars simply migrate onto the top of the tract on the HR diagram, where stars become white dwarfs.

If enough mass is added to the core, so that it reaches 0.44 solar masses, the helium core will ignite and what remains of the hydrogen envelope will shrink back onto the hot, helium-burning core. The star will not be a red giant, but instead be a hot horizontal branch star, white or blue in color, falling into

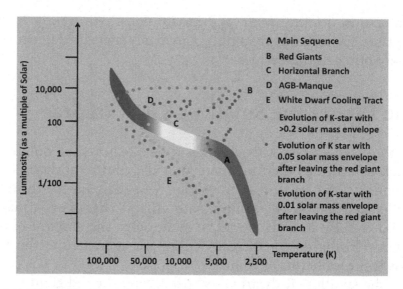

FIG. 3.7 The evolution of K-class stars with differing masses of hydrogen in their envelopes. Those with the lowest envelope masses (•) avoid the return to the red giant branch and simply contract and heat up once the helium is exhausted

spectral classes A or B. These stars have a hot, helium-burning core overlain by a thin, residual shell of hydrogen, which may continue to burn slowly.

Even stars with masses just above the helium-burning limit may traverse the horizontal branch in its entirety until they approach the white dwarf cooling tract. As the hydrogen shell is depleted, the star as whole will contract and heat up. These stars may arrive at the so-called blue hook – the extreme end of the horizontal branch at spectral class O (Figs. 3.7 and 3.8). These hot blue-hook stars retain a residual hydrogen layer with a mass a few times that of Jupiter (~1/200 to 1/1,000th of a solar mass) that is still burning atop the degenerate helium core. These hot stars cut it close.

As these shrivel, remnant stars descend the white dwarf cooling track from ~85,000 to 30,000 K, hydrogen burning is just sufficient enough to raise the mass and temperature of the degenerate helium core to the point at which the helium ignites off-center. Explosive helium burning lifts the degeneracy and the core begins to convect. Any remaining hydrogen is then drawn inwards from the thin, residual envelope, and the star begins a phase of

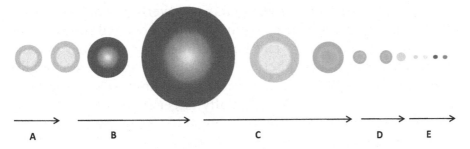

A   B   C   D   E

FIG. 3.8 The later evolution of a 0.55 solar mass star (on the M-dwarf/ K-dwarf boundary). At the end of the star's main sequence life (A) the helium-rich core contracts and the outer envelope expands – the star becomes a red giant (B), perhaps 50–100 times its original diameter. Stellar winds intensify during this phase, removing most of the envelope (C) and leaving a progressively smaller, hotter star that evolves left across the HR diagram and onto the blue-Hook (D). For a while helium is then burned to carbon and oxygen (D), but so little hydrogen remains that the star slips quietly onto the white dwarf cooling tract (E) as soon as the fuel is exhausted

vigorous helium burning. This is accompanied by rapid fusion of the hydrogen that remains in its envelope.

The structure of these stars is rather complex, even though they only have a helium core and hydrogen-rich envelope. Their small girth and very steep temperature gradient drives a complex mesh of interacting zones of nuclear fusion. Remember these stars are only a few times the diameter of Earth – not much bigger than a white dwarf. Their core boils at 100 million Kelvin, while their surface has a temperature of perhaps 30,000 K. That's a big difference in temperature over a very small distance.

These ~0.46–0.49 solar mass stars spend about 1,000 years undergoing the helium flash accompanied by one, then two, overlying zones in which hydrogen is burning. Modeling suggested that initially, during the peak of the flash, all three zones of fusion are mixed together in a very complex mess. However, as the flash dies away, convection, driven by helium burning, retreats as well, and three separate shells emerge. The innermost convection zone is associated with continued helium fusion; this is overlain by another convective shell that is strongly enriched in carbon, where hydrogen burns through the CNO cycle. Finally this shell is capped by a final convective region, where hydrogen burns more leisurely through the *pp*-chain. This outer burning zone extends convection to near the surface of the star a few 100 km further out.

Once the amount of hydrogen in the overlying shells becomes depleted, the hydrogen burning shells, and associated zones of convection, then die away. The helium-burning zone then extends outwards through a number of minor flashes, and the star finally stabilizes on the horizontal branch. The luminosity for stars on the extreme blue hook is marginally larger than the present Sun (~1.18 times) but with a surface temperature exceeding 30,000 K. Life for these stars is nearly over.

Pulsations

Stars evolving from the red giant branch across spectral classes A through F (white through yellow-white) will encounter instability and begin to pulsate. Pulsations are driven by cycles of ionization and recombination of helium ions with electrons at temperatures around 50,000 K. As helium heats up, one of its two electrons absorbs a photon and jumps off the atom, leaving it ionized. This process absorbs energy, trapping it within the outer layers of the star. Consequently, the star swells up, causing it to brighten. As the star swells, the density of the outer layer declines along with its temperature, which allows the trapped energy to escape. Once this happens, the star is able to relax and contract. The process then repeats as contraction increases the density of the layer, trapping energy once more. Such white giant stars are called RR Lyrae variables, after the prototype, and many are known in the ancient, metal-poor globular clusters.

Pulsations in RR Lyrae variables typically occur over a period of a day or so, and the process is relatively well understood. Pulsations have in some cases been used to study the interior of these stars. Astroseismology uses the frequency of the pulsation to determine what structures lie below the stellar surface.

RR Lyrae stars and other helium-burning low mass stars form an extended group, stretching from roughly mid-way down the red giant branch towards the white dwarf cooling track at luminosities roughly 100 times that of the Sun. The majority of these stars have relatively thick hydrogen envelopes and will then evolve back towards the red giant branch as their helium core is exhausted and begins to contract. These stars then form the asymptotic giant branch (AGB) stars that were described earlier. However stars with lesser envelopes of hydrogen are unable to do this and evolve differently. Regardless of their final mass on the horizontal branch, all of the evolving K-class stars will ultimately cross the instability strip and become RR-Lyrae stars for at least part of their later life. The more massive ones will cross it to the left when igniting helium and then once more after the helium runs out. The least massive will make one transit on their way to the blue hook.

The End of the Lowest Mass K-Dwarfs

For stars with marginally more mass – perhaps 0.52–0.55 times the mass of the Sun – once helium fusion ends the star begins a complex death dive. Situated just above the horizontal branch is a diffuse region of fairly hot, luminous stars that have completed helium fusion in their hearts (Fig. 3.9). Although those stars that retained a fairly massive hydrogen shell returned to the red giant

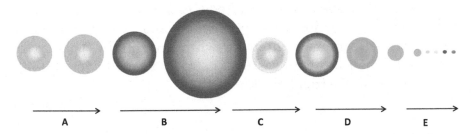

Fɪɢ. 3.9 The evolution of a slightly more massive 0.6 solar mass K-dwarf that retains more than 0.52–0.55 solar masses of gas while a red giant. As in the previous figure, the star leaves the main sequence (A) and expands into a red giant (B). During B and C mass is lost from the outer layers in a strong stellar wind, and the star begins to heat up. During C the star burns helium for 100–150 million years before the core is exhausted. During D the star begins to expand again as its carbon-rich core collapses, and another strong wind sets in. This disperses even more of the remaining hydrogen-rich gas. However, where the envelope is thin, most of the radiation from the core escapes directly, and the star enters the AGB-Manque phase (D). During D the star meanders lazily up the HR diagram until all the available fuel is spent. Once the thin hydrogen-rich envelope has blown away the star becomes a white dwarf (E). There is unlikely to be enough hydrogen left in the outer layer for the star to form a planetary nebula

branch as AGB stars and the lowest mass blue-hook stars simply expired and passed onto the white dwarf tract (Fig. 3.7), those with a mass of hydrogen between these two extremes follow a different path.

These transitional stars, with sufficient reserves of hydrogen, can trap enough of their internal energy in the envelope, causing the star to inflate. However, the hydrogen shell is too mediocre to trap enough energy to drive the star to become a red giant. Instead, the star brightens at nearly constant temperature, ascending the HR diagram in a manner equivalent to a drunk meandering along a sidewalk (Fig. 3.7). As the last of its fuel is consumed in waves of helium and hydrogen fusion the star zigzags up the HR diagram above the horizontal branch. These stars are known as "AGB-Manqué" stars. With insufficient hydrogen and helium remaining, these stars will not affect the glory of current planetary nebulae. Instead they will simply contract and drift onto the white dwarf cooling track once their supply of hydrogen is exhausted or blown off into space. In the process these stars heat up once more, to temperatures of around 80,000 K before shriveling away as white dwarfs.

Like other spent stars, the odd lucky cooling ember, one which had slightly more helium, may undergo a late helium flash, igniting this reserve as it enters the white dwarf track. That exception aside, their ultimate destiny is simply to chill over 300–400 billion years into a frigid blackness.

Conclusions

From our perspective stellar evolution plays the biggest role in the habitability of any surrounding worlds. It is certainly the most dramatic of players, turning stars from dwarfs to giants and back again, while steadily transforming their composition from hydrogen to helium and beyond. Dramatic though these changes are to stars, ultimately they turn out not to have the biggest impact in the long-term habitability of planets orbiting them.

Both the M- and K-dwarfs have lifetimes that we now realize are likely to extend well beyond the capacity of their planets to maintain habitability. Even the most massive K-dwarf may outlive its planets by 10 billion years. It is to the causes of this discrepancy that the next two chapters turn. Within the M- and K-dwarfs the stellar engine may be long-lived, but the fires operating within the reach of their planets have lesser strength, and it is these that will condemn the life on their surfaces to a briefer reign.

4. The Living Planet

Introduction

Early on the morning of Boxing Day, 2004, the day following Christmas, a portion of Earth's crust nearly 1,000 km long broke off the coast of Indonesia. One side of the fracture bounced up 15 m while the other slipped down towards Earth's interior along a fault known as a megathrust. Nearly 300,000 people lost their lives in countries spanning the circumference of the Indian Ocean in the calamity that followed. The accompanying 9.2 magnitude quake was one of the biggest ever recorded, and it shook the planet like a bell. Dramatic though this quake was, it was far from unusual and formed but a small paragraph in a single chapter of one book in the library of our planet. The Boxing Day tragedy represented a small shift in the positions of parts of the planet's surface – a process, called plate tectonics – that may have operated since our planet's crust solidified.

Although its consequences are often catastrophic for us, the process of plate tectonics is utterly essential for the stability of both the planet's interior and atmosphere. It may even contribute to the formation of a stable magnetic field around a planet, a shield that helps ward off much of the harmful radiation from the Sun and other celestial bodies. In this chapter we examine the process of plate tectonics and scrutinize its possible role in generating and maintaining a stable biosphere. The presence of active plate tectonics may determine the long-term habitability of other worlds.

Plate Tectonics: A Primer

Plate tectonics is a natural outcome of an active, living planet with a hot, fluid interior. Fluid is a relative term. In Earth's interior it takes tens of millions of years for material to traverse the 2,000

D.S. Stevenson, *Under a Crimson Sun: Prospects for Life in a Red Dwarf System*, Astronomers' Universe, DOI 10.1007/978-1-4614-8133-1_4,
© Springer Science+Business Media New York 2013

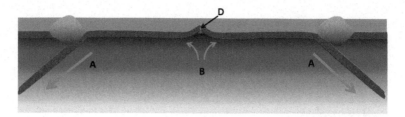

FIG. 4.1 A plate tectonics primer. At *A*, dense, cold oceanic crust plunges into the mantle. About 35 km or more beneath the surface, basaltic crust transforms to the dense rock eclogite. This rock pulls the upstream crust downwards. At *B* the effect of crust moving apart allows the mantle to bulge upwards and partly melt. This forms basaltic magmas (*D*) that rise upwards and solidify to produce new oceanic crust. This forms the undersea chain of volcanoes known as the mid-ocean ridges, or rises. In ocean basins, such as the Atlantic, where subduction is minimal, the crust moves slowly at 1–2 cm per year. In the Pacific, which is fringed by subduction zones, plates move at a nifty 15 cm per year. This illustrates the pulling power of subduction

or so kilometer depth of the mantle. Material moves at a rate of a few centimeters per year driven by differences in the density of hot and cold rock.

The mantle isn't really fluid in the sense that we commonly experience. Instead it behaves more like Silly Putty™. Leave a blob of this curious material lying on a surface, and after a few hours it will have oozed outwards into a thin and rather messy pancake. However, whack it with a hammer and it will shatter like glass into an even messier configuration that takes hours to clean up. Under hot, high pressure conditions the apparently solid rock like deforms over millennia, flowing upwards when hotter and less dense, and then sinking back when colder and denser.

Figures 4.1 and 4.2 summarize these ideas. Plate tectonics is then a superficial expression of convection within Earth's mantle. The hot interior of Earth is trying to cool down by redistributing energy created or trapped within. Inside the rocks of the mantle, radioactive elements, left over from the formation of the Solar System, generate heat as they decay into more stable isotopes. These are primarily isotopes of uranium, thorium and potassium. Together with heat released from the slowly cooling core, this keeps the mantle hot.

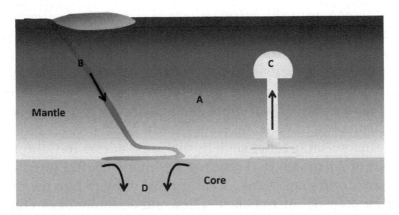

FIG. 4.2 Heating and cooling in the mantle and core. In the mantle (A) the decay of radioactive isotopes heats the mantle rock. Subduction of cold ocean crust drags cold material through the mantle, cooling it (B). At C material at the base of the mantle is heated strongly by the hot iron core underneath. This becomes buoyant and rises upwards in large plumes. These heat the mantle as they rise. Subducted cold ocean crust probably also drives convection in the core by cooling down its hot molten metal (D)

All of that heat has to go somewhere. Where the mass of the body is low, heat is lost rapidly through radiation from its surface into space, and the interior solidifies from the outside inwards. Once a sufficient mass of the body has solidified, heat is lost through this solid layer by conduction. However, if the body is sufficiently large conduction is inefficient, and heat has to be shed in different ways. Instead of conduction the material begins to flow as though it were a liquid, much like water in a pan on a lit stove.

Given that within Earth the core broils at nearly 6,000 °C and the mantle above has temperatures in excess of 1,200 °C, there is plenty of heat to get rid of. Convection within the mantle, in turn, puts stress and strain upon the solid crust above, causing it to bump and grind. This is plate tectonics in a nutshell. Although the details of how the mantle convects are still controversial, no one doubts that as a whole the movement of the plates above is an expression of what happens below. In essence plate tectonics is a surface manifestation of a hot Earth that is trying to cool down. Through the constant churning action within the mantle, heat is redistributed from inside to outside.

Consequently, the entire surface of our planet shuffles around at rates varying from a fraction of a centimeter to as much as 15 cm per year. Motion within the mantle subdivides Earth's surface into provinces of various sizes. In the gaps between some of these plates, hot magma rises upwards and cools onto the plate edges, building up chains of volcanoes and undersea ridges. In other areas the crust is fractured, and pieces slowly grind past each other in long, sinuous fault zones such as the San Andreas in California or the North Anatolian in Turkey. Near many coastal provinces of the Pacific or Indian Ocean, oceanic crust plunges into the mantle, creating deep ocean trenches. These subduction zones consume cold ocean crust, recycling much of it back into Earth's interior.

It is now clear that the motion of cold, subducted crust into and through the mantle is the dominant process that chills it. With that in mind on might presume that the mid-ocean ridges represent the areas where hot material is convecting upwards from far below, replenishing the crust that is lost to the depths through subduction. Yet, as H. L. Menken would observe, a simple answer for a complex question is almost invariably wrong. Until a couple of decades ago most textbooks would indeed have shown hot material convecting upwards under the ocean ridges, spreading outwards and cooling, before sinking back down at the deep ocean trenches.

However, it now appears that the ocean ridges are largely passive players in the process of plate tectonics. Instead of driving plate motion, they are instead formed in response to pulling forces acting on the overlying crust. As plates diverge in response to the dragging power of subduction zones, for example, the crust is thinned and opened up. This reduces pressure on the underlying mantle, causing it to bulge upwards and partly melt. Melting under the ridges is a response to the drop in pressure, which would otherwise keep the rock entirely solid. Melting at mid-ocean ridges liberates the magmas that form new ocean crust atop the mantle bulge. These rocks are exclusively basalts and are relatively dense, dry and poor in the mineral silica (silicon dioxide).

Punctuating this picture of whole crustal motion are hot spots. These appear to be plumes of very hot material that rise up from the base of the mantle. The suspicion is that these carry heat away from the core, dumping it into the mantle and into the crust above. Figure 4.2 shows how hot spots fit into the current picture of plate tectonics.

Glorious Granite

If basalt was the only rock produced by volcanic activity the world would be both a more boring place and perhaps one less well suited to the development of intelligent, complex life. However, plate tectonics has given Earth another rock, one that has had a profound influence on the geological and biological history of our planet – granite. One can argue quite successfully that story of granite is the story of intelligent life (Table 4.1).

Granite is a wonderful rock, an attractive, hard material famed for its use in building and in home furnishings. It's also a fairly simple rock, made primarily of three silicon and oxygen-rich minerals (silicates): feldspar, mica and quartz. These are accompanied by an adage of rarer minerals, such as amphibole and tourmaline. In addition to silicon and oxygen there are smaller quantities of potassium, sodium, calcium and aluminum. Unlike basalts, iron and magnesium are rare in this rock. This makes granite a low density rock, rising upwards, raising its head both figuratively and literally above the waters. Granite sits high in the mantle, allowing it to rise above the surface of our oceans, forming the bedrock for much of the complex animal and plant life that dominates our world. Without granite we might not have warm-blooded mammals. Why is this?

Oceans are probably not the place to develop warm-blooded, mammalian life. For one thing, water has what is known as a high specific heat capacity. This means that water can absorb a lot of

Table 4.1 The density of continental crust, oceanic crust and underlying mantle rocks

Rock type	Density (g/cm³)	Main minerals
Continental crust (granite)	2.6–2.7	Feldspar, mica and quartz
Oceanic crust (basalt)	3.0–3.3 (increasing with age)	Feldspar, pyroxene and smaller amounts of olivine
Eclogite (metamorphosed basalt)	3.5	Garnet and pyroxene
Mantle (peridotite in upper mantle)	3.3–5.7 (increasing with depth)	Olivine, pyroxene with smaller amounts of feldspar

heat energy without warming up substantially. Thus to maintain its warmth a mammal has various adaptations that generate more energy and conserve it. It would seem unlikely that such adaptations would emerge with any frequency in an aquatic environment. Terra firma is probably a pre-requisite for the development of warm-blooded life. The requirement of mammals to be warm internally means that they have to waste additional energy from food to maintain this condition. This requires the oxygen-consuming process: aerobic respiration. The cold oceans are simply too deficient in oxygen to support the kind of metabolic activity needed to power a large brain, or at least develop one. Air breathing organisms access air that is 21 % oxygen; aquatic organisms suffice with only 9–12 %. Thus mammals that live in the oceans or in rivers are compelled to extract oxygen directly from the air.

Although many fish dwarf their warm-blooded, mammalian partners in size – the seals and walruses – they are restricted in their activities, with low levels of metabolism and a limited brain capacity. If we want air-breathing organisms with warm blood, we probably require warm, dry terra firma on which to develop and to evolve. The oceans are no fit environment for driving the development of high-maintenance, warm-blooded mammals. On Earth all of the aquatic mammals developed first on dry land before evolving to forms that could live in the oceans. It is thus to glorious granite that we likely turn our attention if we wish to find intelligent life, as it is granite that builds the greatest proportion of dry land on our world. On a planet with less water, other rocks may suffice, but Earth, and possibly many other planets, will need a foundation of granite on which to develop the most intelligent species.

Given its ubiquity on Earth, granite is an unusual rock, since cosmologically, it is intrinsically rare. This is particularly true if we consider the abundance of basalt in other astronomical bodies. Iron and magnesium-rich basalts are found ubiquitously in the local, and presumably distant, universe. Basalts form the floors of our oceans, along with lunar maria, the Martian volcanoes and much of the surface of Mercury, asteroids and many of the lavas of Io. Basalt is everywhere: granite is not. Why this dichotomy? (Fig. 4.3)

A reasonable solution to this conundrum lies with the process of distillation. Consider the synthesis of Italy's deadly drink, grappa. Grappa is powerful stuff, with 35–50 % alcohol, while its parent drink, wine, has typically 5–15 % of this compound. Grappa

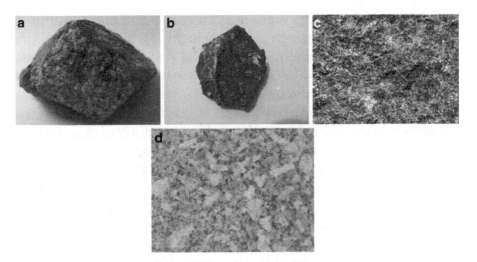

FIG. 4.3 **Key players in the formation and persistence of continents.** (a) Basalt from oceanic crust, seen here with green olivine crystals. (b) Serpentinite – the mantle rock, peridotite, is transformed into this when it comes into contact with seawater. (c) Eclogite, with its rich mixture of red garnet and green pyroxene, is produced when basalt is subducted into the mantle. (d) Finally, glorious granite, the bedrock of continents. The white minerals forming the bulk of the rock are feldspars – calcium, sodium and potassium-rich silicate minerals. The black mineral is mica with smaller amounts of an orangish mineral, amphibole. The remainder of the rock is made of a sparkling clear mineral, quartz. Quartz is pure silicon dioxide. This indicates that the rock was formed through an intricate line of production involving several rounds of distillation from the more magnesium and iron-rich parental, mantle rock. Rocks **a**, **b** and **d** are from Cornwall, England, wile rock **c** is from northwest Scotland

is distilled – a process that concentrates the alcohol in the drink and removes much of the water. When grappa is made, the mush of fermenting grape pulp is sieved (or fractionated) to remove the solid material. The leftover mixture is heated until the alcohol is driven off. These alcoholic vapors are collected, forming the final, more alcohol-rich drink.

Likewise, the compound silicon dioxide is found throughout the mantle, though in a far smaller proportion. Through a complex series of distillations, first basaltic and then granite is extracted from the mantle. However, the analogy falters somewhat here: granite is not so much distilled but brewed out of the mantle by hot water. A better analogy still is provided by the icy drinks known as slushies. After a brief look at the key players, the rocks, the slushy analogy should help explain the process through which granite is born.

The upper mantle is made of a dense, iron and magnesium-rich rock called peridotite, which contains roughly 35 % of the compound silicon dioxide, with the majority of the remainder being magnesium, calcium and aluminum. Peridotite is an even denser rock than the basalts that form the ocean floor. To put things in perspective basalt contains approximately 45–52 % silicon dioxide, while granite may contain up to 75 % silicon dioxide, much of it in its pure form, quartz. The processes known as partial melting, together with another known as fractionation, explains these differences.

If you want to understand the process of partial melting, buy a slushy. Slushies are intrinsically annoying. You buy a frozen solution of flavorings and color in ice. However, as you walk away from the vender and start to drink it, the process of partial melting extracts its cost. Steadily, the more enticing, flavored part is extracted, leaving purer – and blander – ice behind. The flavoring comes out of the icy mix preferentially. The more sugary, colorful compounds cause some of the ice to melt by raising its melting point and then partition, or move, into this melt. Therefore, once the slushy starts to melt all of the fun colors and flavorings move into the liquid fraction, leaving less and less in the frozen portion. Hence your initially flavorsome slushy becomes blander and blander until it becomes tasteless. Likewise, within the mantle, the presence of hot water and carbon dioxide removes potassium, sodium, calcium and silicon dioxide from the bulk mantle rock when it approaches its melting point. Consequently, instead of getting a pure melt of the whole rock, a fraction of it is melted. This melt is the fraction that is most easily extracted in water at lower temperatures (around 800–1,000 °C).

This hot solution (molten rock) is less dense than the surrounding mantle rock and is driven out of the mantle much like tea, liberated from a bag of dried leaves. The molten fraction then rises into the overlying crust. Although much of this chemical broth forms basalts, further rounds of distillation can occur, forming much more silica-rich rocks, including granite. Moreover, as the melt rises and cools down, chemical compounds with the highest melting points begin to crystallize out. These tend to be the denser, iron-rich minerals that are depleted in silicon dioxide (silica). This process is known as fractionation. As these magnesium-rich minerals freeze out and collect near the base of the crust, the very light, silica-rich rock, granite, is produced.

Its low density causes it to rise high into the crust, forming highlands. Thus granites form much more elevated regions relative to the denser basaltic crust of the oceans.

The formation of granite depends exclusively on the presence of water. Without it the processes leading to its formation cannot occur. Within our Solar System granite has only been confirmed to exist on Earth. It is suspected of forming some or much of the highlands on Venus, on which oceans may have existed a billion and more years ago. There is no clear evidence of granite on Mars, any asteroids, or the volcanically active moon Io. In order to synthesize granite, copious amounts of water must make their way into the hot, underlying mantle. It is this prerequisite that makes plate tectonics both necessary and sufficient to produce granite. Only plate tectonics delivers enough water from the oceans above to Earth's mantle below. To understand how the oceans help produce the continents we must return to plate tectonics.

The Mantle and Oceanic Crust

Granite has its origin within the dense, hot rocks of the mantle. The upper mantle is made of a rock called peridotite. This dense mantle rock is a mixture of three main minerals: primarily a dense, iron-rich, green mineral called olivine; another less dense mineral, rich in iron and magnesium, called pyroxene; and much less abundantly a third lighter mineral called plagioclase feldspar. Feldspar is comprised mostly of the elements aluminum, silicon, oxygen and calcium, as well as varying amounts of potassium and sodium. There are a few more minor minerals in this rock as well, but these make up a tiny fraction of the rock and can usefully be ignored here.

Basaltic oceanic crust is approximately 6 km thick. Its lowermost layers are made up of basalt's more crystalline siblings, dolerite and gabbro. These are organized into distinct bands according to where they formed within the crust. The top layers consist of sediments and basaltic lavas, and underneath these are the more crystalline rocks, dolerite and finally, at the base, gabbro. Beneath the crust is the denser zone of frozen mantle rock. This frozen mantle layer, coupled to the overlying basaltic crustal rock, is called the lithosphere.

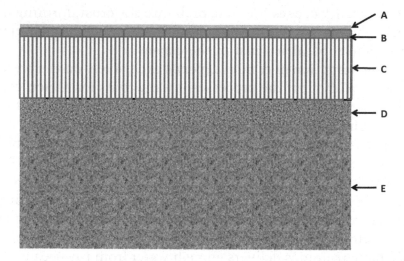

FIG. 4.4 The generalized structure of oceanic crust and lithosphere. Layer *A* is sediments; *B* is basalt lavas; *C* is a thicker layers of dykes – frozen channels of rock that fed the lavas above; *D* is a layer of coarsely crystalline rock called gabbro; and underlying the whole crust is a zone (*E*) of frozen mantle (thicker than shown here) made of peridotite and other olivine-rich rocks

The upper basaltic layer and much of the more crystalline basement rocks are chemically altered by water from the oceans above. Much of the olivine is turned into the greenish mineral serpentine, and much of the other iron-rich minerals become the green mineral, amphibole. These "wet minerals" are important in the genesis of granite (Fig. 4.4).

Why, then, does the melting of peridotite form the lighter rock, basalt? The answer takes us back to our slushy. Under the crust and frozen upper mantle, the hot peridotite is close to its melting point, but the high pressure keeps it largely solid. As the mantle bulges upwards under the mid-ocean ridges, its pressure is reduced, and the peridotite begins to melt. However, as you may now expect, peridotite doesn't melt completely. Instead anywhere between 1 % and 25 % of the rock becomes fluid; this fluid is primarily the lighter minerals pyroxene and plagioclase feldspar. The majority of the denser olivine stays behind. Melting forms a liquid portion that separates from the denser, olivine-rich fraction under gravity. The product rock is our basalt, which then rises up through the cracks at the ridge, forming new oceanic crust.

Dotted along the length of the ocean ridges are areas of high heat flow. Here water, circulating within the hot rock, drags super-heated, mineral-rich waters upwards and out into the cold, dark ocean floor. These torrid flows produce remarkable undersea kingdoms of life, dominated not by sunlight but by the chemistry of the deep Earth. Indeed there is a substantial, if not universally accepted, body of evidence that roots the tree of all living organisms on Earth to these hot undersea worlds.

At a very fundamental level, plate tectonics links the geological heart of the planet to the living, breathing biosphere above. Furthermore, the discovery of these life forms had profound implications for our concept of the "habitable zone." Before the discovery of these undersea kingdoms in the early 1980s it was assumed that all life depended on sunlight for energy. Thus a habitable world would be a planetary body orbiting its star at a distance that provided sufficient warmth and energy to maintain liquid surface water. However, the discovery of life that lives independently of our star means that worlds that have liquid water buried under kilometers of rock or ice may well be just as habitable for life as ours. Admittedly, that life is likely to be microbial, but it may be just as evolved as anything on Earth.

Eclogite

Returning to our oceanic crust, how does this new rock age and return to the mantle? When the ocean lithosphere is young it is warm and marginally less dense than the hot, underlying mantle. However, as the lithosphere ages, it cools down, thickens and its density increases.

When the ocean lithosphere is perhaps 180 million years old, the lithosphere is so cold that it is now denser than the underlying hot mantle. A fatal and irreversible process then begins. As the crust becomes denser it begins to sag further and further into the mantle. This eventually tears it away from any associated continental rocks nearby. The dense ocean lithosphere then begins a journey downwards into the mantle under its own weight. The key factor in the motion of oceanic plates and the formation of continents is a very attractive red and green rock called eclogite (Fig. 4.3). The role of eclogite has additional profound implications for the activity of plate tectonics on alien worlds. However, at birth oceanic crust has none of this rock, so where does it come from?

To answer this question, we must look at the life of a basalt that has erupted on the ocean floor. As we have seen the basalt erupted at ocean ridges are strongly altered, or metamorphosed, through their interaction with seawater. The water converts the basalt to a rock called amphibolite. The iron-rich rocks take up water and trap it for the remainder of the amphibolite's life. This water is the key to the formation of granite.

When the oceanic crust ends its life and plunges back into the mantle at subduction zones, the water is brewed out of the crust and enters the overlying mantle in the subduction zone. After the water has been released, the altered basalt (amphibolite) continues to be compressed and heated until the minerals within it rearrange. Inside Earth this process occurs at depths of 35 km or more, where suitable pressures, and temperatures in excess of 600 °C, are reached. The aluminum and calcium-rich silicates of the parent basalt are transformed into the luscious red garnet, while the blackish pyroxene of the basalt is retained in a different, contrasting greenish form.

Eclogite is considerably denser than the neighboring mantle rocks at an equivalent depth. Consequently, once this tipping point is reached the now denser oceanic crust is pulled downwards into the depths of the planet. This has the effect of dragging the trailing crust along as well. Subduction then becomes irreversible. If one looks around the Pacific Ocean the majority of the basin is fringed by subduction zones. It is no coincidence that the fastest moving crust on Earth is found in the Pacific Basin. The combined pulling power of all those subduction zones is dragging the crust away from the spreading centers where the crust is produced and back down into the depths of the planet.

The process of eclogite formation has important ramifications for the operation of plate tectonics on other, more massive planets. The transformation is pressure- as well as temperature-dependent. On a more massive super-terran – the type of planet most commonly observed to date within the habitable zones of red dwarfs – eclogite formation will occur at much shallower depths, conceivably at the base of the oceanic crust because the force of gravity is greater. On such worlds, plate tectonics may not operate, at least not in the style we might be familiar with. Oceanic crust might form and founder in chaotic patches, forming small, unstable plates. Interestingly, within early Earth, and on planets with hot interiors, this instability might lead to the rapid formation of continental crust. The surface of such a planet may be dotted with patches of continental crust, jiggling around between small plates or patches of oceanic crust, rather than be subdivided into the primarily large territories seen on Earth.

Constructing Continents

Although subduction marks the death of oceanic crust, it means the genesis of new continental crust (Fig. 4.5).

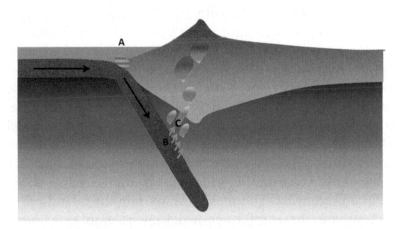

FIG. 4.5 The formation of continental crust. When basaltic ocean crust and upper mantle ages it cools and becomes denser. After a 180 million years or so it becomes denser than the underlying mantle and sinks into it. At A, thick layers of sediment is scraped off, and some is dragged down into the mantle, where it melts. At B the wet ocean crust bakes dry, and the water released rises into the mantle wedge, C, lowering its melting point and causing it to fuse. The rising basalt that is produced cools and begins to crystallize. Heavy iron-rich minerals settle out, leaving lighter granitic magma behind. This rises up into the crust. Some erupts as lava, while most crystallizes underground, forming new crust. The melting can also directly form granite when the rock that melts does so at a lower temperature (partial melting)

Continents are born in the fire of destruction. As the water-logged oceanic crust dives into the mantle and begins to transform into eclogite, the water is released, and the process of granite formation begins. The partial melting of mantle rock liberates the magmas that initially form lines of volcanoes above the descending limb of the neighboring oceanic plate.

Volcanic activity builds chains of islands dominated by basaltic volcanism called island arcs. These arcs are consistently found 150 km or so behind the line of subduction. The distance is dictated by the depth to which the ocean crust has descended before it is dehydrated. The location of the overlying island arc is therefore a reflection of the process of subduction.

Initially basalt dominates the erupted lavas. This is hardly surprising, as basalt is the rock produced by partial melting of the mantle. However, as these island arcs grow higher and the crust underlying them becomes thicker, smaller and smaller quantities of basaltic magmas make it to the surface. These basalts tend to become trapped and cool within the thickening overlying crust. Time spent underground affords the basalts time to crystallize. As the denser, iron-rich minerals crystallize and fall to the base of the thickening crust, the less dense, silica-rich material is left behind, forming increasingly large island masses. Significantly, these low density islands cannot be subducted into the mantle – they are simply too light. When plates bearing island arcs collide, the islands remain bobbing above the surface of the waves, even as huge masses of oceanic crust tear away into the hot mantle underneath.

Over time collisions produce larger amalgamations of land, much like present-day Indonesia. Many famous island arc volcanoes dominate these micro-continental assemblages, including Pinatubo and Krakatau.

Assembling a Continental Jigsaw

A continent has a humble beginning. It is a journey stretching from inception at the creation of garnet within a subducting slab of basalt, through to the amalgamation of island chains into a resolutely persistent continent. The process may take billions of years.

Despite its extended history some particularly large continents can assemble over shorter intervals of time. A particularly spectacular example occurred during the Phanaerozoic era – from 543 to 250 million years ago. During this period a large number of small continental fragments were milling around near the equator, north of what was then the supercontinent, Gonwanaland. Flanking many of these disparate fragments were long lines of subduction. Throughout this period of time, subduction generated large belts of island arcs riddled with active volcanoes.

Large masses of newly synthesized granites were born beneath these volcanoes. As the ocean crust between the lines of subduction shrank, these belts of islands crashed together. Swept up among these were large masses of ocean sediment, dead organisms and other ocean detritus. As these collisions continued what is now China and the newly formed continent of Asia collided, forming lines of new mountains. Finally, around 300 million years ago, Europe completed the sequence, colliding with the expanding Asian continent and throwing up the Ural Mountains. From nothing more than small fragments of land and crust sweated from the mantle, a new continent was born: Eurasia. Stamping its authority on the planet Eurasia survived the subsequent fission of Pangea and now forms the largest block of continental crust found on Earth today.

Like the island arcs that came before, continents cannot be subducted. Once formed, that's it. Imagine trying to push a block of polystyrene into water. You can shove it down but it pops straight back up again when you release the pressure. In essence this is what happened off the coast of Indonesia in 2004. The Indo-Australian oceanic plate was pushing down under the continental crust underlying Sumatra. However, because of the immense frictional forces acting against it, the down-going oceanic crust was dragging the overlying continent with it. The buoyant continental crust was trying its best to resist the downward push, and something had to give. After decades of being dragged downwards, locked to its descending oceanic partner, a portion of the continental crust broke free of its oceanic chains, bouncing back upwards 15 m along a 1,000 km stretch of fault. Simultaneously, the oceanic crust slipped downwards towards the mantle. The rapid upward thrust displaced millions of tons of seawater, generating the tsunami that killed so many.

Japan bore witness to the same process in March 2011. These events illuminate a process that has been shaping our world for billions of years and is responsible for the creation of the land on which we live. Such human tragedies sadly mark the synthesis of new continental crust. Earth is slowly brewing new crust onto which our descendants may 1 day live.

The Role of Mantle Plumes in Continent Formation

Essential though subduction is, it is not the only manufacturer of new crust. Today, mantle plumes play an important role in the formation of oceanic islands and probably also influenced the synthesis of early continental crust. When plume meets oceanic crust, the crust bulges upwards, and an enormous series of underwater eruptions commence. As the ascending plume of rock rises it decompresses. Decompression lowers the melting point of the rock just as it does under the mid-ocean ridges. Melting produces copious amounts of basalt – proportionately more than is synthesized at the mid-ocean ridges as the temperature of the plume tends to be higher than found under the ridge. Higher temperatures allow more melting. This is particularly true when the head of the plume first arrives at the base of the crust. Large undersea eruptions can build extensive submerged plateaus known as provinces, the largest known of which is the Otang-Java plateau in the Western Pacific. When these large plateaus arrive at subduction zones they become stuck and eventually have their bulk fused onto any nearby continent. Through this route, mantle plumes can help forge continents and may have been important in this regard in Earth's past.

Dotted around the globe are the fossilized remains of earlier plume arrivals. In Scotland and Northern Ireland a large igneous province harbors some of the most beautiful landscapes in Western Europe. These include such rugged vistas as the Quiraing on the Isle of Skye, used in the film *Prometheus*, and the famous Giant's Causeway in County Antrim, Northern Ireland. In Africa, the Karroo basalts form areas of highland in Angola and Namibia, while in India the Deccan Traps form a large plateau in the west of the country.

After the plume head has dumped its mass of heat into the crust, a train of partially molten rock continues to rise upwards, forming the tail of the plume. These tails provide prolonged episodes of volcanic activity that can be traced across ocean basins and continents alike. In the Pacific, the most famous of these is currently building the Hawaiian Islands. The plume pushes the crust upwards, and supplies a fairly constant amount of basalt, progressively building up islands. Meanwhile the constant northwestern motion of the Pacific plate carries the islands away from the plume towards the Aleutian subduction zone. The islands and underlying crust steadily cools, contracts and slumps back down, as the volcanic activity dies away. One by one the islands erode and slip away into the ocean depths, only to be replaced by new ones arising at the new southeasterly end of the volcanic chain.

In the Atlantic the existence of Iceland is a consequence of the same plume that formed the basalts of western Scotland. Both Iceland and, much further south, Tristan da Cunha, are plumes that happen to have met the surface along the length of a mid-ocean ridge. For them this means that they can remain in one spot while magma rises underneath. Iceland thus grows and grows while the Hawaiian Islands lead a fleeting existence.

Where a plume meets continental crust, there is a very different outcome. As before the arrival of the plume head often causes the crust to bulge upwards. This has the effect of causing the crust to extensively fracture – often in a so-called triple junction – an arrangement of three radiating areas of fracture, each directed away from the center of the bulge. This can have the effect of helping fracture continents into smaller fragments. The tug of an adjacent subduction zone may be needed to cement the divorce. However, there is ample evidence on Earth that the arrival of a plume-head may be what a continent needs to break apart and in the process generate a new ocean.

However, trapped beneath the continent, these normally passive basaltic lavas can drive much more explosive events. The origin of these lies in the melting of the silica-rich continental crust. If a continent winds its way over a plume the hot basaltic rick may melt its way into the overlying granitic crust. This produces very silica-rich magma that can force itself through the crust. Silica-rich magmas are far more viscous than basalts. These often trap gases, leading to violent explosions. Recent examples of this violence can be found in Yellowstone National Park in Wyoming, where a simmering super-volcano lurks under the pristine waters of Yellowstone Lake.

There is also good evidence that plumes may also be the source of our most precious jewels – diamonds. When a hot spot arrives at the base of a continent, it heats the overlying lithosphere, the frozen keel of mantle rock that is welded to the base of the continent. In some instances the plume or lithosphere may contain large quantities of carbon dioxide. Under the immense pressures found within the lithosphere, the carbon in this gas, as well as carbon present in the mantle, may transform into diamonds. Where there is sufficient gaseous carbon dioxide, the hot mantle plume then begins a rapid, explosive puncturing of the overlying crust. A hot, gas-rich blast rips open the crust and drives nature's most resilient mineral into the upper crust; shortly thereafter the shock wave sends gases violently into the atmosphere. The peculiar magmas formed are kimberlites. These magmas freeze rapidly within the crust, forming deep vents filled with a mess of different mantle and crustal rocks. Hidden within this morass of debris are the occasional gemstones that drive humans to tunnel far into the ground in search of wealth.

On early Earth such acrimony must have been rare. The amount of surface covered by continent was limited. Most of Earth's surface was covered in thin oceanic crust. However, as the planet has aged, the abundance of thick continental crust has increased, making the impact of plumes on plate motions more effective. Continents are fairly easy to push around. They may not be subducted, but they do respond readily to movements in the mantle around or underneath them. When a critical mass of continental crust is produced, the surface dynamic of the planet alters, and new ocean crust can be made, divisively, within continents as well as alongside nearby continents. Continents then become players in the game of life and death that punctuates the history of the oceans.

Since continental crust is intrinsically buoyant, it constantly builds up on the planet's surface. Once it is made it will last as long as the planet does. Contrastingly, oceanic crust has a finite lifetime. Created at an ocean ridge, it slowly moves outwards, cooling, thickening and slumping into increasingly deep water.

Ultimately, the entire crustal thickness is peeled away, dipping downwards into the hot mantle below. Oceanic crust has a limited lifespan.

On Earth, there is evidence that the process of plate tectonics has been active for at least 3 billion years. When subduction happens and fragments of continental crust collide small fragments of oceanic crust, called ophiolites, become trapped in among the continental wreckage. These can be found in among all modern mountain chains such as the Alps or Himalayas. Fossilized ophiolites can be found elsewhere, such as the heavily eroded, ancient highlands of Scotland, or the Appalachians in the eastern United States. Each of these key terrains is a fingerprint of mountain building and serves notice on the foreclosure of an ancient ocean basin.

Just How Does Plate Tectonics Work?

Plate tectonics is regarded as a theory. This is an awkward and unfortunate position, at least in terms of public perception. To the general populace, the word "theory" suggests that scientists are unclear as to whether plate tectonics actually happens and consequently that ultimately it may simply be wrong. Clearly plate tectonics does work. Plate motion can be measured with some precision using the GPS satellite networks obsessed over by map-illiterate drivers.

The use of the word "theory" lies not with the process as a whole but with the minutiae of the mechanisms that drive it. There is considerable controversy as the extent of mantle convection. Conflicting geophysical and geochemical models show convection spanning the entire breadth of Earth's mantle, or confined to distinct layers, and this impacts on our understanding of alien worlds, too. The picture of our planet's interior is very cloudy indeed, an image that has recently been muddied further by laboratory experiments. Yet that should not detract from our revolutionary understanding that the surface of the planet is mobile.

The reason for all this conflict stems from contrasting and often seemingly incompatible observations. Seismology can be used to probe the interior, while lavas erupting at the surface can bring chemical clues from the depths. Earthquakes are sound waves that pass at different speeds through materials with different chemistry and density. When an earthquake occurs two main types of wave – the fast moving back-and-forth P-wave and a slower, side-to-side S-wave spread outwards. P-waves travel through solids and liquids with relative ease. The denser the material the wave is passing through, the faster it travels. S-waves will pass through solids but not liquids.

As these waves spread through the mantle, waves passing through columns of hotter, less dense rock will slow down, and those passing through dense, subducting oceanic lithosphere will travel faster. By measuring the time it takes different waves to move from source to receivers around the world, geophysicists can construct a portrait of the planet's interior in much the same way as a CAT scan can provide a detailed map of a patient's body. This is called seismic tomography. Seismic tomography has visualized subducting oceanic plates passing through the full depth of the mantle, thus mixing both upper and lower mantle in one vast convection cell.

Yet seismological observations are full of contradictions. For some considerable time seismology has made it clear that the mantle is broadly divided in two. The top 660 km consists primarily of the dense, olivine-rich rock, peridotite. Throughout this region, the density steadily increases until the silicate minerals comprising this rock break down and reassemble into new forms. At around 660 km depth a much denser rock, perovskite, emerges, along with smaller proportions of similarly dense silicates and metal oxides. These are then believed to comprise the bulk of the lower mantle. The passage of seismic waves indicates a clear demarcation of both layers. But does this schism indicate that there is a restriction in the flow of material between each layer in the mantle?

Thus we have a situation where seismic tomography suggests a pattern of mantle circulation opposing that seen with other seismic surveys. It also appears to conflict with evidence from lavas (geochemistry) And while tomography shows that some slabs are deflected when they reach the boundary between the upper and lower mantle, most travel through unhindered. This is a difficult nut to crack.

While geophysicists prefer whole mantle convection, geo-chemists point to hot-spot magmas that appear to tap material that is primordial in nature and has ascended from the deep inte-rior. Gases found within hot-spot magmas also appear to show evidence of untapped deep mantle. Yet many hot-spot magmas also appear to contain material from subducted oceanic crust. The picture is very messy indeed.

Finally, in 2012, sophisticated high-pressure experiments indicated that the lower mantle is more silica-rich than the upper mantle, apparently confirming earlier suspicions that the two mantle sections are distinct. Preservation of such a discontinuity would imply that the upper and lower mantle don't mix much even over billions of years of Earth history. How can this finding be reconciled with the passage of subducted plates into the deep interior?

Perhaps, one could arrive at the same end point (a silica-enriched lower mantle) through a more prosaic process that we already know occurs. Oceanic crust consists of basalts, which have higher silica contents than the underlying mantle. Subduc-tion should thus enrich the lower mantle in material from the more silica-rich basaltic crust.

Over Earth's history the continuous delivery of more silica-rich basalt, or more accurately, eclogite, from subducting oceanic crust to the lower mantle might just produce the observed dis-continuity in silica content. The mantle is slowly turning over prolonged periods as cold, dense, but the more silica-rich rock sub-ducts and piles up in the lower mantle, while the less dense but more mafic (iron and magnesium-rich) residue is slowly displaced upwards. Chemical processes, as well as the punctuated passage of plumes through this mess, both complicates this picture and drags material upwards into the upper mantle.

The nature of the mantle is not an idle concern. Whether the mantle convects as a whole or is split has profound implica-tions for how planets transfer internal heat. In turn this deter-mines how extraterrestrial worlds will evolve as they age. In more massive super-terrans the pressures rise faster internally, so the peridotite-perovskite transition happens much closer to the plan-et's surface. If mantle convection is stifled by this transformation, plate tectonics could, in principle, be inoperable on more massive super-terrans. However, this effect is confounded by two other processes: the formation of eclogite and the effect of increased gravity on the strength of the lithosphere. A greater gravitational field increases mantle pressures at any given depth, compared to Earth. Therefore, basaltic crust could founder earlier if the base of the crust converts to eclogite. This is likely to be more important early on when magma production is greatest and the crust correspondingly thicker. Premature formation of eclogite would lead to a different pattern of plate motion but might not preclude it.

Work by Diana Valencia (then at Harvard University) examined the effect of an increased gravitational field on the ability of plate tectonics to operate on more massive worlds. Calculations by Valencia and colleagues suggest that the higher gravitational field leads to a thinner lithosphere (crust and solid uppermost mantle) and to more effective subduction. Since subduction seems easier on super-terrans and is the key process driving plate tectonics, then it appears as though plate tectonics is inevitable on more massive, terrestrial worlds. Diana Valencia's work even suggested that dry super-terran planets, with little or no surface water, will still have plate tectonics. On the less massive Earth, ironically, many calculations suggest that our planet should lack surface tectonics. It appears that water is an essential component of the process on our planet, but this may not be true of all planets. It will be interesting to see whether any dry super-terrans show evidence of active plate tectonics – though obviously observing the process, or its signature, on worlds trillions of kilometers away would be a challenge in itself.

Paul Tackley's (Institute of Geophysics, Zurich) and Diana Valencia's (MIT) groups also re-examined the effect of the perovskite transition on mantle convection in super-terrans with masses up to ten times that of Earth. Their work indicated that despite the formation of perovskite, and at higher masses even denser states called post-perovskites, mantle convection still operated. The pattern of up- and down-wellings altered with mass, but convection still transferred matter from the top to the base of the mantle and back again in even super-terrans with ten times our planet's mass. Thus it would appear that a very massive, rocky super-terran can still operate this most essential process – plate tectonics.

If we assume a lack of plate tectonics, this would in turn limit the exchange of gases between the mantle and atmosphere, which would have profound implications for the long-term survival of any life on such a world. The resolution of this dilemma seems as far away now as it did 20 years ago. Geochemists and geophysicists seem locked in a battle over how the interior of our planet is mobilized. If we can't discern the truth for our world, imagine the ramifications for planets tens of light years away!

The Past, Present and Future of Tectonics on Earth and Other Worlds

Geologists are clear that plate tectonics has operated on Earth for at least 3 billion years. The evidence is found in the bands of preserved oceanic rock known as ophiolites. However, it may not have operated in the earliest epochs of the planet's existence. For one, recent studies suggest that there has been a fundamental change in the rate of continent formation from 3 billion years ago – suggesting a different mechanism operated in the more distant past. Continent formation requires the hydration of oceanic crust, and this is dependent on the depth of the oceans. If the oceans were much shallower in the past than they are at present, continent formation would not have proceeded in the manner it does today. Finally, modern plate tectonics requires the formation of eclogite at depth within the mantle. Early in the planet's history, higher mantle temperatures may have retarded the creation of this most essential rock.

Geologists have thus suggested two fundamentally different modes of both plate tectonics and continental crust formation that may have occurred early in the planet's history. Soon after the planet formed the hot upper mantle would have melted much more extensively than it does now. Perhaps as much as 50 % of the rock would melt at temperatures in excess of 1,600 °C. This is far greater than the 1,250 °C maximum seen with the formation of basaltic lavas today. The key rock produced at high levels of partial melting is not basalt. It is an extinct species of Earthly rock called komatiite. Komatiites are much denser than basalts, containing far less silica and far more iron and particularly magnesium. The rock is rich in both the minerals olivine and pyroxene and relatively deficient in basalt's key mineral, plagioclase feldspar. Created at ancient mid-ocean ridges, komatiite could, in principle, drive plate tectonics, much like eclogite does today. Given that its density is much like the mantle underneath, soon after its synthesis, strong cooling would allow its density to rise sufficiently to drive it back down into the mantle beneath.

However, there is considerable debate as to the extent to which komatiites substituted for basalts. Many of the planet's fossilized komatiites appear rich in potassium and may be associated with subduction – the destruction of oceanic crust – rather than with its creation. Other komatiites are clearly associated with mantle plumes. As the plume head arrived, basalts were erupted in profusion. Later, as the hotter tail of the plume reached the surface, komatiites were erupted. The youngest of these plume komatiites were erupted in the Mesozoic era in the Caribbean and are well documented.

The association with mid-ocean ridges is far more equivocal. Therefore, these dense, iron-rich rocks may not be the sought-after solution to the operation of plate tectonics in the past, but instead be associated with its consequences.

These problems aside, a komatiite ocean crust opens up a possible interesting solution for the formation of the first continents and may therefore be relevant to the operation of plate tectonics and the formation of continents on alien worlds. Work by Tomas Naeraa and colleagues (Geological Survey of Denmark and Greenland) suggested that there were key differences in the processes that formed continental crust at the beginning of the Archean compared with later in that era and subsequently. Their work suggests that prior to 3.2 billion years ago the early crust was formed by simple re-working of older material. Analysis of isotopes of the rare metal hafnium suggested that crust was rapidly recycled through an unknown mechanism. After 3.2 billion years ago, the signature of subduction – eclogite – emerges in the isotopic record, hidden in the continental crust. Crust formed after this date is clearly derived from the mantle. Earlier crust seems to be derived from some reservoir of material with a chondritic composition, i.e., not affected by differentiation. The most likely source is a mass of primitive crustal material that is simply re-melted.

One model for the formation of early continental crust involves the re-melting of over-thickened oceanic crust. This would put it in line with Naeraa's work. One key rock type characteristic of early continental crust but largely absent today is found as TTG terrains. TTG stands for the three principle silica-rich rocks contained: tonalite; trondhjemite and granodiorite. These are granite-like rocks but with higher quantities of sodium than is seen in modern granites. The oldest continental crust belongs to this series. The high-sodium content argues for their formation via partial melting of eclogite rather than the wet partial melting of peridotite seen today. This process simply does not happen in today's mantle because the temperatures within it are too low to melt eclogite at suitable pressures. However, TTGs could have formed in the past if the mantle was particularly hot, or if the crust was much thicker. This would allow eclogite to form directly at the base of the crust, then melt, forming the TTGs.

How could the thick crust needed to form the TTGs have come about? More than a decade ago, in the early 1990s Maarten J. de Wit (University of Cape Town) and Andrew Hynes (McGill University) had suggested that the source of this reworked crust was komatiitic oceanic crust that was imbricated (Fig. 6.6). Imbrication simply means crust that is broken into a number of slices that are piled up on one another.

The model goes something like this: Prior to 4.2 billion years ago any oceans would have been shallow and the mid-ocean ridges may have stood proud of the ocean surface. However, with continued out-gassing of the mantle, and comets and asteroids bringing more water, they continued to fill. Comets might have delivered much of this water during the late heavy bombardment, 3.9–3.8 billion years ago. As the oceans filled these komatiite ridges would have submerged. The resulting hydrated rock is not sufficiently dense to subduct, but will instead pile up upon collision. These deep piles of thinly interwoven rock are the imbricates. As these piled higher their bases would have pushed down into the hot mantle, transformed into eclogite and melted. The product of all this melting would have first been basalt and then, as the piles of basalt grew thicker, the sodium-rich granites – TTGs. This style of tectonics would eventually cease once basalts came to dominate the composition of the ocean crust and normal subduction commenced – perhaps, as the record seems to indicate, around 3.2 billion years ago.

Indeed, studies of preserved ancient continental crust by C. Brenhin Keller and Blair Schoene (Princeton University) suggested that early continental crust was formed in a manner dependent on the presence of unusually thick oceanic crust. This could have come about through the piling up, or imbrications, of wet komatiite-rich crust.

The early continents could also form in a second, distinct way involving mantle plumes. In today's world large, hot mantle plumes release copious amounts of basaltic lava when they impact the crust above. Presumably in the past, a much hotter interior would accommodate a greater number of these plumes, and the volume of lava released with each arrival would have been proportionately greater still. In 1992 A. Kroner and P. W. Layer (Institute for Geowwissenchaften) suggested that the first continents may have formed from large oceanic plateaus dotted with volcanoes. Imagine a world populated not by continental terrains but by large islands analogous to Iceland. As magmas ascended through them, and as hot, underlying mantle cooked their bases, these undersea kingdoms of rock produced the TTG cores of the modern-day continental crust.

It is important to stress that prior to 2.5 billion years ago – the point at which the Archean era gives way to the Proterozoic – very high heat flow could have allowed far greater extents of partial melting within the upper mantle. Not only could this have produced komatiites but also should have generated a much thicker oceanic crust in general. During the Archean era, the base of this thick crust would have reached its melting point, and the TTG continental crust would have been produced. In today's mantle temperatures are too low, and continental crust can only form where water is introduced by subduction. As the melting point of the mantle rock was now lower than before, the composition of the rock was then changed from sodium-rich to the present potassium-rich form (Fig. 4.6).

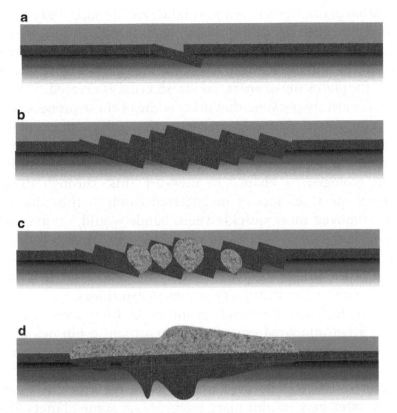

FIG. 4.6 One scenario for the formation of the earliest continental crust in the Archean era or earlier during the Hadean. In *A* wet, komatiite-rich oceanic crust is too light to subduct, so builds up into thick piles (*B*) when it collides. In *C* the base of these imbricated piles becomes dense and hot enough to turn to eclogite and then melt. Melting of eclogite liberates sodium-rich granites. These are lighter than the komatiites and so rise upwards, forming the first buoyant continental crust (*D*). Delamination of the lower eclogite part of the lithosphere will have the same effect if it melts as it plunges into the mantle

The Organization of Early Plate Tectonics

The early Earth, being far hotter than present, had to dump its excess heat somehow. There are a number of ways in which this could be accommodated. Firstly, there could be more mantle plumes arriving at the base of the crust, dropping off their heat as they do so. Secondly, plates could move more quickly. A faster rate of creation of oceanic crust and subduction of more cold crust into the mantle could keep the mantle cool. Finally, a hotter mantle could generate more ocean ridges than are found on the modern Earth today. A greater length of ocean ridge allows more heat to be lost as the plates move apart, and fresh crust is created.

It is entirely possible that all or some of these processes could have been at work, but at least as far as the geological record is concerned, plate motions certainly weren't much faster 3 billion years ago than are seen today. Fossilized oceanic crust – ophiolites – affords geologists a chance to measure this, through measurements of the thickness of magnetized bands within the basalt. If plates moved more quickly these bands would, on average, be broader, but they aren't. Early continents also pay testament to plate motions not much niftier than today. The solution to the release of the planet's early heat may lie primarily with plumes and a greater rate of heat loss through ocean ridges.

If one looks at the torrid Jovian moon, Io, the very high heat flow is accommodated entirely through mantle plumes and tidally generated fracturing of the crust. Based on their high temperatures, at least some of the lavas that erupted on Io appear to be komatiites. There is no evidence for plate tectonics. Although a lack of water may inhibit plate tectonics on some planets, it may also simply be that very hot interiors lose heat mostly through mantle plumes rather than plate motion. The early Earth may thus resemble a super-terran planet with heat loss dominated by plume activity rather than plate tectonics. There could, therefore, be interesting consequences for the development of life, as the next chapter explores.

What, though, is the fate of plate tectonics on a planet? Can it continue indefinitely, and if not, how might the planet behave as its interior cools down?

The Construction of Continents and the Heat-Death of a Planet

Continental crust is produced by the process of subduction – where old and cold oceanic crust is destroyed new continent is created. In these inventive steps the rocks also concentrate much of the heat-producing radioactive isotopes, such as potassium. Indeed, on Earth nearly half the entire planet's inventory of uranium, potassium and thorium has been extracted from the bulky mantle and dumped into the much less imposing continental crust through subduction. By extracting the very materials needed to keep the interior active, the formation of continental crust has foreshortened the active life of our planet's interior.

Thus there is an inherent contradiction in the life of a planet. Plate tectonics distills the ingredients needed to sustain it and brings about a faster rate of cooling of the planet's interior. This is a consequence of both subduction cooling the mantle and the loss of the radionuclides that keep it hot. So, how long can plate tectonics remain operational on a planet? The answer is determined by the mass of the interior and the proportion of radioactive elements the planet was bequeathed at birth.

Some planets will be born with lower budgets of these unstable elements than Earth and consequently will cool faster. However, if we keep this variable controlled the following general rules apply. The abundance of heat-generating elements will scale with planetary mass, and the heat of formation of the planet will also increase likewise. A larger body also has a lower surface area to volume ratio. What does this mean? As the mass increases, the surface area of a sphere increases in proportion to the square of the radius, but the volume increases by the radius to the power of three. This means that the volume goes up faster than the surface area, and larger planets lose heat less effectively than small ones. A larger, denser super-terran might then retain enough heat for plate tectonics to operate for billions more years than it will here on Earth. However, this is not a straightforward relationship.

On Earth plate tectonics has operated in some form for perhaps the planet's entire history, and given a reasonable future rate of cooling will do so for another 500 million years, or so. Cessation of plate tectonics will occur on our planet when its interior has cooled sufficiently for the lithosphere to become so thick and rigid that it is unable to fracture in response to the motion of the mantle rock below. On planets two to three times the mass of Earth, calculations suggest a fairly straightforward scaling, and plate tectonics could operate for 8 or 9 billion years. However, if the surface becomes dominated by granite, the low density of this rock will bring the cycling of rock between the crust and mantle to an end. This process is described fully Chap. 5. This could ensure that the motion of the surface ceases far earlier than it will on Earth. Thus the geological life of a planet will diverge strongly from ours in a manner dependent on its mass. This, in turn, may radically alter the ability of a planet to sustain biological life in the longer term.

Plate Tectonics and the Stability of a Planet's Biosphere

Aside from its role in rearranging landmasses, plate tectonics has two important consequences for the planet as a whole. Plate tectonics delivers cold, dense oceanic crust to the mantle, where it ultimately descends, in many if not all cases, as far as the core-mantle boundary. Where the descending plate reaches this layer it cools the outer, liquid core of Earth. The net effect is to drive convection in our planet's liquid core. Without this cooling effect, radioactivity in the mantle and conductive heating from the core below will warm the mantle up. Without a cooling influence from above, the temperature gradient within the core will then decrease, as less heat is being lost from its upper surface. The implication is that convection would decrease, or even cease, in the planet's liquid outer core. Since convection within the core seems necessary to help establish the planetary magnetic field, this magnetic field may falter or fail. Hence a planet without plate tectonics will lack an appreciable magnetic shield, and its atmosphere will be subjected to a thorough battering from the solar wind.

Despite nearly equivalent sizes and masses, Venus and Earth have divergent histories. At some point in its past Venus lost its magnetic field. Although the timing is not well constrained it may have been coupled to the loss of surface water and the consequent cessation of plate tectonics. As a result the atmosphere of Venus is now being stripped away by the solar wind. A trail of gas streams outwards past the orbit of Earth. At some point in the future, the dry, hot Venus may lose the bulk of its atmosphere and end up like Mars, another world lying exposed to the full wrath of the Sun.

Conclusions

In many senses discussing plate tectonics in a book dedicated to the discovery of life away from Earth may seem a tad abstract. However, without plate tectonics three essential ingredients for life may be lacking: a stable biosphere with sufficient carbon dioxide to support life; a crust sufficiently thick to host complex life; and a stable magnetic field. Plate tectonics appears to be the mechanism that interweaves all three of these ingredients. An active, living interior is almost certainly the key piece in the development of complex life. The star provides the long-term stability of global temperature and energy for biological processes, but the internal engine stabilizes the effects of the central star and maintains habitability in ways that may be more subtle but are equally profound.

Therefore, it falls to the process of plate tectonics to maintain long-term habitability. Plate tectonics has a finite life, governed by the amount of heat a planet has and by how quickly its surface clogs up with granite. Should these processes end early the other systems on which life relies will fail. It may well be that on a crimson world it falls to plate tectonics to deliver the final kiss of death to life, rather than the star itself.

5. The Carbon Dioxide Connection

Introduction

Carbon dioxide is regarded as a dirty chemical at present. The molecule conjures up images of melting ice caps, flooded cities, and animals and plants pushed to the edge of extinction. However, recent bad press hides carbon dioxide's fundamental role in creating living worlds. Carbon dioxide is both a protective blanket and nurturer of complex life. In this chapter we explore its role in the physiology of a living world and its complex relationship with both living organisms and the geology of the host planet.

Carbon Dioxide: The Essential Gas

Every high school student should be able to tell you that carbon dioxide is a raw material for photosynthesis. It is the carbon scaffold around which more exciting and tasty molecules, like glucose, are assembled by plants and bacteria. Without carbon dioxide, the starting point for carbon's great chemical journey through our living organisms is lost and life becomes impossible. Plants and microbes are the foods of animals. Take away carbon dioxide and you take away plants; take away plants and you remove the animals as well.

However, for life on Earth, and possibly elsewhere, carbon dioxide is something of a double-edged sword. Animals respond very badly to carbon dioxide concentrations above 5 % of breathable air. The gas dissolves in their bloodstream and turns into carbonic acid and its derived salts. This triggers the nervous system to accelerate heart and breathing rate, fearing that the blood will become too acidic or that the level of oxygen has fallen too low.

D.S. Stevenson, *Under a Crimson Sun: Prospects for Life in a Red Dwarf System*, Astronomers' Universe, DOI 10.1007/978-1-4614-8133-1_5,
© Springer Science+Business Media New York 2013

Eventually, such an acceleration in both heart and breathing rate leads to cardiac arrest. The essential gas of photosynthesis poisons the animal.

Unsurprisingly plants, lacking a respiratory system, have no such worries and positively thrive at higher carbon dioxide concentrations than are currently found on Earth. For them higher carbon dioxide means more fuel for the fire of photosynthesis, and their growth rate goes up.

Admittedly, this assertion is something of an oversimplification. There are many types of plants that utilize carbon dioxide with differing efficiencies. Use of carbon dioxide at higher concentrations than are currently found on Earth also requires that every other potential variable for growth is optimized. These include temperature, the availability of nutrients and the amount of oxygen that is present. Having a perfect inventory is rarely the case, as has been made clear in experiments investigating the effects of higher carbon dioxide on crop yield on Earth. Yet, in general, we can say that given the opportunity most plants do better with more carbon dioxide.

The Fate of Carbon Dioxide

Carbon dioxide is a reasonably abundant gas. This is a reflection of the abundance of the two elements that comprise it. Carbon and oxygen constitute the third and fourth most abundant in the universe. This proportion is dictated by the vagaries of the nuclear reactions within the stars that formed them. In planets, carbon may be delivered as carbon dioxide ices, their associated salts or as more complex organic compounds. Inside the bulk of an accreting planet, organic carbon will be heated strongly with silicate-rich rocks; oxygen reacts with these compounds to form carbon dioxide. The volatile carbon dioxide is then boiled upwards until it is released into the atmosphere by volcanoes, or through cracks in the surface. Assuming the temperature of the atmosphere remains reasonably low (less than a couple of 1,000°) and the amount of hydrogen is limited, carbon dioxide simply accumulates (Fig. 5.1).

However, carbon dioxide is vulnerable to assault from water. If the atmosphere is cool enough for water to condense, carbon

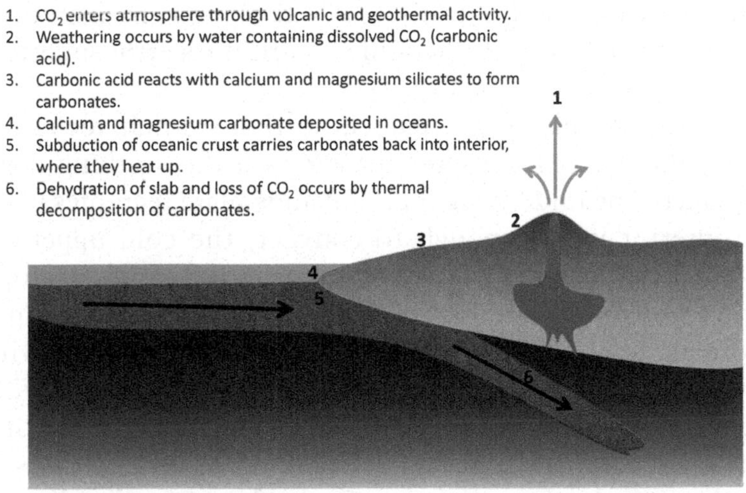

1. CO_2 enters atmosphere through volcanic and geothermal activity.
2. Weathering occurs by water containing dissolved CO_2 (carbonic acid).
3. Carbonic acid reacts with calcium and magnesium silicates to form carbonates.
4. Calcium and magnesium carbonate deposited in oceans.
5. Subduction of oceanic crust carries carbonates back into interior, where they heat up.
6. Dehydration of slab and loss of CO_2 occurs by thermal decomposition of carbonates.

FIG. 5.1 A simplified diagram showing the carbonate-silicate cycle. CO_2 cycles back and forth from mantle to atmosphere, driven by mantle convection. *1.* CO_2 enters atmosphere through volcanic and geothermal activity. *2.* Weathering occurs by water containing dissolved CO_2 (carbonic acid). *3.* Carbonic acid reacts with calcium and magnesium silicates to form carbonates. *4.* Calcium and magnesium carbonate deposited in oceans. *5.* Subduction of oceanic crust carries carbonates back into interior, where they heat up. *6.* Dehydration of slab and loss of CO_2 occurs by thermal decomposition of carbonates

dioxide dissolves in it, forming a weak acid called carbonic acid. The same acid carves the magnificent limestone caves such as those found in Mexico, Malaysia or Yorkshire, England. Once it has dissolved in water, carbon dioxide can still escape through the reverse of the chemical reaction that stole it from the air in the first place. Hydrogen carbonate can break down into carbon dioxide and water.

Yet, once carbon dioxide is locked into hydrogen carbonate, it risks further reactions with calcium and magnesium in silicate rocks. This reaction forms the insoluble compounds calcium carbonate (chalk) or magnesium carbonate (dolomite). Once in this form, it tumbles along river systems before sliding relentlessly to the bottom of any ocean. Many aquatic organisms also grab hold of it, forming protective shells. This carbonate also winds its way to the floor of any ocean once the organism dies. Thus, even without the grabbing hands of photosynthesis, carbon dioxide's shelf-life within the atmosphere is limited. Bereft of a mechanism to return it, any world rich in water would soon find its atmospheric tanks of carbon dioxide running dry.

On geologically active planets, carbonate-rich rocks eventually find themselves transported back into the seething depths of their planet. The underlying mechanism is plate tectonics (Chap. 4). If the interior is hot enough to convect, the cold upper surface of the mantle becomes too dense to remain floating on top of it. After a few tens of millions of years, the denser crust slips back into the planetary interior, dragging any sediment laden with carbonates along with it. After millions of years this crust will have descended several tens of kilometers into the planetary interior and be rapidly warming up. As the temperature approaches 1,000° the carbonates decompose, fracturing back into carbon dioxide and oxides of calcium and magnesium. Free once more, carbon dioxide meanders its way through the dense mantle rocks and back out into the atmosphere, through fissures or volcanic vents. The cycle begins anew.

On a terrestrial planet such as Earth, the key factors in the maintenance of this cycle are a hot interior and a wet surface. Water not only cools the crust above, but as it descends with the underlying crust into the mantle it depresses the melting point of the hot mantle rocks, causing them to melt at relatively low temperatures. This liberates magmas that rise upwards, bringing with them the essential gas (Chap. 4).

Without water the melting point of the mantle rocks is far higher. A drier mantle is thus more viscous and the crust more rigid. Without water a planet's crust will lock in place, much like that of present-day, bone-dry Venus. Without water the rigid Venusian crust cannot subduct. Moreover, the crust is simply too hot. This makes it too buoyant. Thus Venus is hit by a double-whammy: a dry, rigid crust that is also insufficiently dense to re-enter the planet's mantle. Any material deposited on the planetary surface cannot be consumed by the mantle and be recycled. Moreover, without water carbon dioxide cannot be fixed either in solution or as carbonate rock. It is locked into the atmosphere.

There still may be reserves of carbon dioxide in Venus's interior. However, sooner or later volcanoes will drain it, leaving all of the gas trapped in the hot, dry Venusian atmosphere (Fig. 5.2).

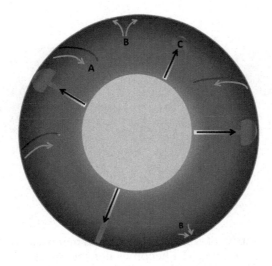

FIG. 5.2 The probable types of convective movement within the Earth's mantle. At *A*, cold and dense oceanic crust sinks back into the mantle and descends haphazardly all the way to the core-mantle boundary. This cools the mantle as it descends. At *B*, the mantle passively upwells beneath the areas of extension, under either the ocean or continental crust, forming areas of rifting or mid-ocean ridges. At *C*, hot cylindrical plumes of rock ascend from near the core-mantle boundary, carrying heat away from the core and delivering it to the top of the mantle

On Earth, the leisurely cooling of the interior has led to a slowdown in the rate of plate tectonics and volcanism as a whole. In the past, plates either moved faster or the total amount of plate edges through which new crust was forming was greater. This was necessary if the planet was to shed the heat from its formation and from radioactive decay (Chap. 4). As the interior cooled mantle convection became more sluggish, and the amount of volcanic activity declined. Thus the rate at which carbon dioxide enters the atmosphere also declined.

Furthermore, plate tectonics has another Achille's heel when it comes to the essential gas. As water enters the mantle at subduction zones, it produces the stuff of continents: granite (Chap. 4) (Fig. 5.3).

With an intrinsically low density, granite does not subduct. What the planet creates it retains. The denser oceanic crust merrily subducts when it exceeds its sell-by date. After 180 million years or so oceanic crust is so dense it is readily returned to the hot mantle below, where it is destroyed (Fig. 5.4).

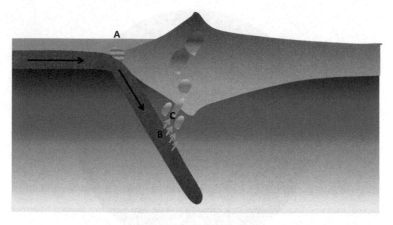

FIG. 5.3 The formation of continental crust. When basaltic ocean crust and upper mantle ages it cools and becomes denser. After a 100 million years or so it becomes denser than the underlying mantle and sinks into it. At A, thick layers of sediment scrape off, and some of it is dragged down into the mantle, where it can melt. At B the wet ocean crust bakes dry and the water released rises into the mantle wedge, C, causing it to melt. The rising basalt produced cools and begins to crystallize. Heavy iron-rich minerals settle out, leaving lighter granitic magma behind. This rises up into the crust. Some erupts as lava, while most crystallizes underground, forming new crust. The melting can also directly form granite when the rock that melts to form it does so at lower temperatures (partial melting)

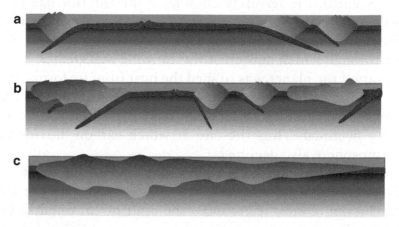

FIG. 5.4 The formation of a thick lid. In (a) plate tectonics generates continental crust through subduction. Early continental crust is a small part of the crust as a whole. In (b) much of the surface is covered by continental crust, and more is being added through continued subduction. By (c) the majority of the surface is covered by continental crust and subduction is now impossible, as the continental crust is too buoyant. There is no further cycling of carbon dioxide between the atmosphere and the mantle

However, the granite has no such attraction to the planet's interior. Granites formed 3.9 billion years ago are still found on Earth's surface. Even when continental crust is attached to a thick keel of frozen, dense mantle rock, its overall density remains so low that the continental crust remains on the surface even when the dense keel peels off into the hotter mantle below. Although plate tectonics recycles the ocean crust, to and from the mantle, the continental crust simply shuffles passively about, pulled this way and that way by convection in the underlying mantle and the drag of any attached oceanic crust. Granite simply basks on the surface of the planet, growing steadily in size as subduction adds volcanic material to its edges. Any carbon dioxide locked up as carbonate rocks on its surface will remain trapped there and be unable to recycle.

Moreover, given time continents can grow so that their edges meet – not through collision but by simple addition of new rock, extracted from the mantle. Therefore, once the continental crust covers a sufficient portion of a planet's surface the continents effectively lock in place and will no longer move around. Plate tectonics is on the ropes. Decreasing areas of oceanic crust can still subduct and recycle carbonate and other materials into the interior, but as their area goes down the efficiency of this process must decrease. In the end the oceanic crust will be confined to small, land-locked basins much like the Black Sea or part of the Caspian Sea. Subduction ceases and the basin simply fills in with sediment.

Unless hot mantle rock forces its way up through the rock layer, carbonates that are trapped on the planet-wide continent will never bake enough to decompose once more. The carbon dioxide will be trapped for good.

As Earth has evolved, more and more granite has been sweated out of the mantle, encasing more and more of the surface with the recalcitrant rock. Consequently, increasing amounts of carbon dioxide have remained fixed to its surface as a thick shell of calcium and magnesium carbonate. Although these carbonates form many beautiful landscapes in their own right, they are a dead end for the essential gas.

Ironically, water thus poses a long-term problem for life. On the one hand it is necessary for life to start, but over geological time it causes the loss of carbon dioxide from the planetary atmosphere, thus inhibiting photosynthesis. On any other Earth-like world this process must happen, too. Carbon dioxide will run out – or at least run down – as long as the surface remains wet.

On super-terran worlds, planets abundant around red dwarfs, this process that forms granite but removes carbon dioxide may be even more taxing. A super-terran will retain a hotter interior for longer than our planet. Conceivably, the process of plate tectonics will occur as well, as we looked at in Chap. 4. However, given a larger mass of mantle to sample, the area covered by continental crust may well rise more quickly than here on Earth. Therefore, the amount of available carbon dioxide might then fall at a faster rate than it has on Earth.

However, there are other reasons why we might be concerned about the long-term habitability of supper-terrans. On a super-terran the higher gravitational pull will tend to cause the hot, soft base of the crust to spread outwards – essentially flowing over the top of any dense, basaltic, oceanic crust. This will tend to clog up the surface faster than it does on Earth. Moreover, a super-terran will be hotter than Earth at any given age because of the much higher heat from its formation and from any radioactive elements contained in its mantle. For this reason volcanism will be higher early on and will tend to produce thicker crust than found on Earth, which will tend to produce more granite and, because of the higher gravitational field, be more prone to spreading than on Earth.

Models by Edwin S. Kite, Michael Manga and Eric Gaidos (University of California at Berkley) suggest that any super-terran with a mass greater than three times that of Earth will likely develop a thick granite lid that chokes off plate tectonics within a few billion years – less than the current age of Earth. Indeed, the crust may become so thick early on that even hot-spot volcanism becomes problematic.

Without effective cycling of carbon dioxide life is in a bit of a fix. Calculations by Manfred Cuntz (University of Texas at Arlington) Werner von Bloh (Potsdam Institute for Climate Impact Research), Peter Schröder (University of Sussex), Christine Bounama and the late Siegfried Franck (both at University of Potsdam) suggest that the longest time a planet may be habitable is around 9–11.5 billion years. The planets with the longest periods of habitability have the smallest amounts of exposed continent. A planet dominated by oceans – possibly all super-terrans above a certain mass – but with 10 % or so continental crust will maintain carbon dioxide levels in a habitable range for around 11.5 billion years.

As the surface area of continents increase, the lifetime of the biosphere decreases, as there will be more land available to react with atmospheric carbon dioxide and remove it from the atmosphere. Conversely, planets with no continental land area (all ocean, or aquaplanets) cannot operate the carbonate-silicate cycle at all, and their habitability will be determined by the rate of outgassing of carbon dioxide from the ocean and the mantle underneath. Both of these rates are difficult to determine and are very sensitive to the temperature of the atmosphere and ocean.

Given these problems is there any way life could persist once a thick lid of granite has formed?

Life Under the Thick Lid

Is the formation of a thick lid the death sentence for life on a super-terran? The answer to this is as yet uncertain, but we can examine some scenarios in which life can continue.

Once the planet-wide granite shell has formed volcanism won't cease, at least not immediately. The interior of a super-terran should be very hot, perhaps several hundred degrees hotter than Earth at any comparable age. All that heat has to go somewhere, and simply trying to lose it through a thick granite shell by conduction may not cut the mustard.

On Earth, the core sheds its heat through the mantle via large plumes of rock. Hot cylinders of rock rise from the core-mantle boundary and ascend the full depth of the mantle, arriving at the top of the mantle at 1,200–1,500 °C. As they approach the surface the pressure on the rock decreases, and copious amounts of basalt magma are produced. These are erupted onto the surface through any fractures present (Fig. 5.5).

Aside from delivering fresh hot magma to the face of Earth, the most important effect of plumes is to push the crust upwards as the plumes approach the surface. In a thick-lid world mantle plumes may be the only mechanism through which carbon

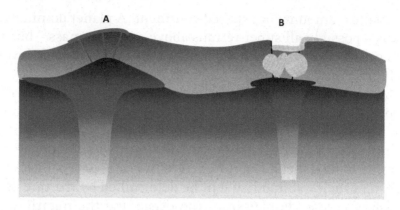

FIG. 5.5 Hot-spot volcanism. Hot plumes of rock rise up from near the core-mantle boundary. When they reach the thick continental crust they may produce abundant basaltic magma, which rises to the surface producing flood basalts (A) such as the Siberian Trap eruptions of 251 million years ago. Or at B the heat from the plume or from the rising basaltic rock can re-melt the thick continental crust, producing large quantities of granitic magma (hatched blobs). This can erupt to the surface explosively, forming large eruptions that produce pyroclastic flows – dense fast moving currents of rock and ash. The Yellowstone National Park caldera and the more ancient Glen Coe in Scotland were produced in such eruptions

dioxide can be recycled. Carbon dioxide can then be delivered to the atmosphere through the volcanism eruptions that accompany their arrival. Further carbon dioxide might be released if the magmas forced their way through rocks rich in carbonates or buried organic material. About 251 million years ago, this appears to have happened when the Siberian Trapps were erupted.

As the crust is stagnant in a thick-lid planet, volcanoes will simply build higher and higher, fed by their mantle plumes. You can see the effect of this on Mars, where Olympus Mons has, over the course of a billion years or so, simply ballooned in size, becoming easily the highest mountain in the Solar System. On Mars the lithosphere is thick because the mantle is relatively cool. However, on a thick-lid planet with a hot interior, the lithosphere may be hotter and melt more readily. This will cause the steady sinking of any mountain, volcanic or other, that is put on top of it. Therefore, on a thick-lid planet, volcanoes will steadily rise, pumping a steady stream of life-giving carbon dioxide gas into the atmosphere. The planet-wide granite lid may have steadily eroded or simply oozed back under a global ocean, but here and there volcanic islands might dot its surface. As before carbon dioxide can react with these fresh volcanic rocks and form carbonates on the flanks of these edifices.

Over time the carbonates will get buried by more and more rock. Eventually these carbonates will become hot enough to break down, liberating their precious store of carbon dioxide once more. Continued eruption, burial and sinking could allow carbon dioxide to cycle into and out of the planet's interior for some time after plate tectonic ceases. Thus even in this extreme world, carbon dioxide may cycle to and from the interior, affording life a chance to continue.

Super-Volcanic Eruptions

There is another, potentially interesting, consequence of hot-spot volcanism on a thick-lid planet. On Earth, mantle plumes interact with continental crust with some frequency. In most cases the result is crustal doming, followed by fracturing, collapse and often extensive basaltic volcanism (Chap. 4). In some cases thousands of cubic kilometers of molten rock are extruded in a very short interval of time. The eruption of the Siberian Trapps was a particularly significant episode in the life of Earth. The effusion of these basalts was tied convincingly to the planet's most severe global extinction event (Fig. 5.6).

However, in the Snake River area of Wyoming and neighboring states, a hot basaltic plume has extensively melted the overlying granite crust, leading to highly explosive eruptions. These culminated in the formation of the Yellowstone caldera a little over 660,000 years ago. On a super-terran, with a thick granite layer, it would seem likely that highly explosive eruptions would be driven by mantle plumes – and these may be the norm. As on Earth large eruptions could lead to periodic mass extinction events.

Plants, too, would periodically struggle if the atmosphere were suitably polluted with debris from these eruptions. However, such eruptions would also replenish the atmosphere with

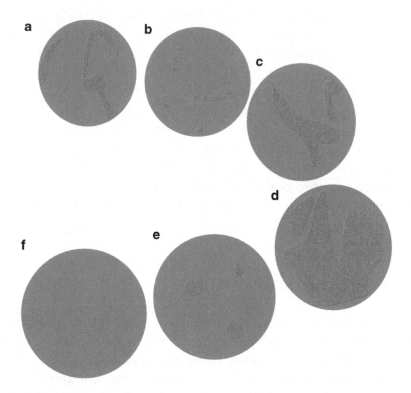

FIG. 5.6 The changing face of a super-Earth. Possible changes to the proportion of surface covered by dry land over eons. At the earliest stages (A)
the oceans are filling and areas of land may protrude above water, such as
mid-ocean ridges. By a few hundred million years after birth (B) the oceans
have filled with their maximum volume, and only islands protrude above
the water. At C plate tectonics has begun, fusing island arcs and other
terrains together to make continents that make their maximum aerial
exposure in D. By a few billion years of age granitic continental crust has
choked off plate tectonics, and the surface becomes locked. Without plate
tectonics mountain building ceases, and the land erodes down until it is
largely submerged (E). Only hot-spot volcanism maintains islands of high
ground. By 5–10 billion years of age, F, hot-spot volcanism has waned, and
all land becomes submerged

carbon dioxide. In a late era when carbon dioxide levels were low
and photosynthesis was struggling, there might be periodic extinction events driven by the lack of carbon dioxide. Super-volcanic
eruptions might push life even closer to extinction by periodically
filling the atmosphere with debris that blocked out light and prevented photosynthesis.

This may sound grim, but it might not be so. Were super-eruptions to occur, these events would be followed by a rekindling of life as carbon dioxide levels were boosted by both the eruption and the decomposition of dying organisms. Indeed, as the amount of carbon dioxide waned, a highly complex process could emerge. Life might teeter on the edge, periodically retreating to the oceans as the abundance of carbon dioxide declined in the atmosphere. Just as the end seemed inevitable a massive explosive eruption could push life's remnants harder, eliminating further species. But as the dust settled, both figuratively and literally, the conditions for life would be restored, and species would make an emboldened return. On an aged super-terran life may cycle from near obliteration to plenitude on geological lifetimes dictated by the ebb and flow of volcanic activity. Indeed, this may also be true of our diminutive neighbor, Mars, where volcanism has ebbed and flowed, seemingly for over a billion years.

Is there a limit to how long the process of hot-spot volcanism can go on? Can the crust become so thick that magma is no longer able to rise through it? This depends on a number of factors. Aside from the thickness of the crust, the composition of both the crust and magma are important. Denser basaltic magma has more difficulty pushing through low density granite crust simply because of the large difference in density. For an analogy, imagine hot water rising through oil – not something that normally happens with ease. The temperature of the magma is also important. The higher the temperature the lower its density and the easier it will find it to push through the overlying granite. Moreover, if it is sufficiently hot it can melt its way through. Volcanism could cease altogether on such planets if the magma is no longer able to penetrate the increasing thickness of low-density granite crust.

At present there simply isn't enough data to give a clear picture of the future of volcanic activity on a planet with a thick crust.

There are two potential means of tackling this question. The first is empirical. We can see that on Mars, which has a thick crust and lithosphere, volcanism has occurred for most of Martian history. Despite Mars' limited mass and lower internal heat, volcanic activity appears geologically recent, if episodic. It may even have occurred as recently as 200 million years ago. Volcanism has persisted despite the 150 km-thick burden of crust through which the dense basaltic magmas have erupted. On Earth, remobilized granite crust produces explosive eruptions. Therefore, even after it becomes impossible for basalts to punch their way through the thick granite layer, explosive eruptions may continue indefinitely as a result of plumes melting the overlying burden of crust. This simply is an unknown at present.

The second solution to this question may come in the future. With improved technology satellites may be able to directly observe volcanic activity occurring on distant super-terrans. Data, most likely, would come in the form of changes to the composition of the atmosphere following large eruptions – such as a measurable increase in the amount of sulfur dioxide. Alternatively, with even greater resolution, we might be able to monitor eruptions through thermal imaging. Where the planets can be aged we can then use this information to constrain the duration of volcanic activity on such worlds. We may be worrying unnecessarily – geology may find a way…

Photosynthetic Miners

Given all these potential troubles, is there another means by which photosynthesizing organisms could get carbon dioxide if the atmospheric supply ran out? Ironically, we may have to turn our attention to the plant's roots rather than its leaves. Plants and many microbes secrete acids into their surroundings as they grow. These acids are primarily derived from the breakdown products of carbohydrates. As every high-school student knows, acids react with carbonates, releasing carbon dioxide. On Earth the amount of carbon dioxide locked up in carbonate rocks is approximately 70 times that found in the atmosphere. Conceivably, if sufficient areas of land stay exposed to air, plants and other living organisms, which critically depend directly on carbon dioxide, could obtain their quotient of the gas by dissolving it directly from limestone and other carbonate rocks. Thus life may indeed find a way even in this difficult environment.

The atmosphere of Earth, 500 million years from now, will be critically depleted in carbon dioxide. Levels will likely fall below 100 parts per million (ppm). Even the resilient grasses face starvation at levels of carbon dioxide much less than 150 ppm. Aside from the dogged cyanobacteria living in the oceans, land plants will be forced to adapt to new sources of carbon dioxide or face extinction. Naturally, if they go, so, too, will all higher life forms that depend upon them. Earth will appear to be running evolution in reverse – plants slowly fading and the landscape turning brown. Life will retreat back into the oceans, where photosynthetic bacteria will persist until the level of carbon dioxide falls short of 10 ppm. However, with abundant surface limestone and other carbonates, plants may turn their roots instead of their leaves into the organs responsible for the acquisition of carbon dioxide. Photosynthetic miners may take over the landscape, slowly eating into it, carving out deep biomes in the crust.

The outcome for life on such a world depends critically on 2 factors. The first is how quickly the continental crust thickens and chokes off volcanic activity, and the second is whether the oceans eventually come to cover the entire planet once plate tectonics has ceased. If the crust thickens early, the land will slip beneath the waters, and effective cycling of carbon dioxide will cease. Moreover, any mining plants will find their habitats drowned out, and what life there is will become confined to the oceans once more. Such organisms will be forced to live directly, or indirectly, from food chains based around bacteria. This is perhaps a grim portrait of life on a super-terran. Despite the seemingly greater prosperity of resources on these larger worlds, geology may play a damning role, condemning life to an early demise. Hence, it may be smaller Earth-like worlds, many of which have now been found around M-class dwarf stars, that are the best hopes for life.

Conclusions

The mass of a planet grossly affects the manner in which it generates and recycles its crust. This in turn has profound implications for the sustenance of any biosphere. Regardless of the initial mass plate tectonics will cease, either because the mantle has cooled and stiffened, or because of the formation of an excessive burden of buoyant continental crust. The timescale of such events appears to be on the order of 5–10 billion years after the planet's formation. This is well short of the life of the star. At present the longest a planet seems able to support life is around 9–12 billion years. And this is for planets with surfaces dominated by water. On planets with more significant amounts of exposed land, erosion of exposed continental crust will have depleted the atmosphere of carbon dioxide by 11 billion years of age at most. This inexorably leads to the end of productive photosynthesis and the termination of all food chains dependent on plants or photosynthesizing bacteria as a source of organic carbon. Aquaplanets may sustain some sort of habitable ocean, but the productive lifetime of this environment is uncertain, given a lack of a carbonate-silicate cycle. Regardless, the formation of continental crust will eventually terminate plate tectonics and doom the aquaplanet to a slow biological death.

Data from the HARPS program has concluded that super-terran worlds are ubiquitous within the habitable zones of red dwarf stars. Several papers were published in 2012 by Xavier Delefosse and Xavier Bonfils that examined the radial velocities of 102 nearby red dwarf stars. In perhaps the most significant of these, statistical analysis of HARPS data suggested that 36 % of nearby red dwarf stars host super-terran mass planets in short period (1–10 day) orbits. A further 35 % of red dwarfs hosted a super-terran in wider 10–100-day orbits. Within this data set, statistics suggested that 41 % of all red dwarfs have a planet lying within their habitable zone. One must say that the range of potential values in the numbers of habitable worlds was substantial. However, the take-home message was clear. Potentially habitable planets orbit red dwarfs within the locality of the Sun – and they appear to be abundant. Thus, in terms of the habitability of worlds, the issue of the longevity of plate tectonics is far from inconsequential. The survival of plate tectonics may play a far bigger role in determining the long-term habitability of the planet than the nature of the host star.

To put this in context let us look, briefly, at the red dwarf Gliese 581d. The planet's host star has a main sequence life of approximately 350 billion years. Plate tectonics and hot-spot volcanism could well have ceased on Gliese 581d, given the presumptive age of the system at 7–11 billion years. Regardless, as far as we are aware geological processes cannot sustain a living biosphere for the full 350 billion years the star will shine. The cessation of volcanic activity will shut down the supply of carbon dioxide long before the star dies.

Although Gliese 581d may be habitable now, there would be considerable problems sustaining the planet's biosphere for much longer. Indeed, if Gliese 581 is 7–11 billion years old, life on this planet may be on its last legs. With an age of 7 billion years plate tectonics may have ceased as much as 3 billion years ago if a thick granite lid developed. Levels of atmospheric carbon dioxide may be so low that complex photosynthetic life may have become extinct, with the only surviving life residing around a dwindling number of undersea hot-spot volcanoes. It may seem sad, but we may well have missed the boat with this world. Its prime years may be gone, with complex life having been resigned to the dustbin of history. All this happened while our planet stumbled through its formative years.

Life on a super-terran may not be sustained despite eons of available stability in the parent star and while the planet has ample internal warmth. It now appears that planetary geology is far more important for the development and stability of life than the star that hosts it. This is all the more true if the star has a very low mass and a correspondingly extended lifespan. Certainly, in terms of sustained habitability, the parsimonious lifestyles of red dwarfs may not be the panacea we once thought. There are far more variables at play than the sustenance of mere sunlight. For the majority of its life, a red dwarf may play host to a retinue of dead planets, orbiting well within its stellar habitable zone. We will return to this in more detail later in the book.

6. Stability of Habitable Atmospheres on Red Dwarf Worlds

Introduction

According to military legend, German gunners encountered a strange phenomenon while attempting to shell Paris in 1918. They aimed their super-gun of the day, the 210 mm Long Max, at Paris from 70 miles to the northeast. It was said that despite a dogged determination on the part of the German military, the projectiles continually missed, curving away and hitting ground 0.7 miles west of the capital. Were the gunners simply poorer shots than they believed, or was something more fundamental at work? This unusual legend illuminates an unexpected property of rotating bodies – the Coriolis effect. This phenomenon has serious implications for the habitability of red dwarf worlds.

Tidal Locking

Planets orbiting within the habitable zones of all red dwarf stars are tidally locked to them, always presenting the same face to the star. This is true for planets orbiting within the habitable zones of the lowest mass orange K-dwarfs as well. Tidal locking means that a planet rotates once on its axis for every orbit of its star. This effect is a result of tides raised by the planet upon the upper layers of their star. The tidal bulge, in turn, pulls upon the spinning planet, gradually breaking its rotation.

The same process happens between Earth and the Moon. Tidal bulges in the oceans, atmosphere and Earth's mantle pulled on the Moon as it orbited Earth. This had the effect of slowing

D.S. Stevenson, *Under a Crimson Sun: Prospects for Life in a Red Dwarf System*, Astronomers' Universe, DOI 10.1007/978-1-4614-8133-1_6, the Moon's rotation, until it spun in synchrony with its orbit, and caused it to migrate further away from Earth, which it is still doing today. With little deviation, the Moon now permanently presents the same face to us. Slight variation in the orbital period of the Moon, and the shape of its barely elliptical orbit and its spin period, mean that the Moon appears to wobble, or librate, as it progresses through its cycle. A planet orbiting a red dwarf may also librate, but effectively one hemisphere will face its sun for all eternity. Meanwhile, the opposing hemisphere will reside under a permanent blanket of darkness.

These un-Earthly conditions will pose considerable challenges for living organisms on these worlds. This comes down to the way the planet receives energy from its parent star and is consequently is heated by it. A planet without an atmosphere would experience strong heating on its sunlit side, while the dark side would freeze indefinitely while its parent star evolved from cradle to grave.

Although clearly not within the habitable zone of our Sun, Mercury encounters a similar phenomenon. On its airless, sunlit side temperatures swelters at more than 350 °C, while the night side experiences temperatures less than–180 °C.

Moreover, organisms that received their energy from the parent star would have to adapt to some uniquely peculiar circumstances if they were going to survive on such a world. The direction in which light was striking the surface of their world would vary in angle with the distance from a location known as the sub-stellar point, or SSP, for all eternity. The SSP is merely the point on the surface of the planet over which the Sun shines directly downwards. Since there is no rotation relative to the star the SSP is fixed. On Earth the equivalent point moves with the rotation of Earth on its axis throughout our day and is always located between the Tropics of Cancer and Capricorn.

The process of tidal locking may take a few tens of millions of years at most for planets orbiting within the habitable zones of the lowest mass stars. Within the planet, during this intense period, vigorous tidal heating may cause violent volcanic activity. It has been suggested by some that this heating could be so severe that it would result in the loss of the hydrosphere – the gaseous and liquid part of a planet – effectively baking the planet dry and thus preventing the subsequent development of life. Even if the planet was thoroughly baked, later collisions with comets and asteroids would most likely restock the planet with enough water to fill its oceans.

The Atmosphere After Tidal Locking

Once tidal locking has been completed the habitable world will settle into a perpetuity marked with some interesting geographical features. Aside from the sub-stellar point (SSP for short) there will be a virtual line in the sand that demarcates the edge of the illuminated and dark hemispheres. This line is called the terminator. This SSP has the sun permanently overhead – permanently meaning for billions upon billions of years, while the terminator has the sun in a state of permanent sunrise or sunset, depending on your perspective. Thus one might intuitively expect gross differences in the temperatures of each hemisphere. However, intuition is often wrong, no more so than in this case. Why is the night side warmer than expected?

In 1998 Martin Heath (the Biospheres Project, London) published some of the first detailed models of the atmospheres of planets orbiting red dwarfs. Heath and colleagues considered both the impact on any life of tidal locking and the types of radiation emitted by these uniquely cool stars. Until this point, many had considered red dwarfs to be poor candidate planets for habitability. With one face locked to the star, the difference in temperature between the sunlit and shaded hemispheres might be so severe that the atmosphere would become unstable. In such a state water vapor and later carbon dioxide would flow from the lit to dark hemispheres, freezing out on the cold, dark side. This process is known as atmospheric collapse.

Although not as dramatic as it initially sounds, atmospheric collapse would have catastrophic effects on any advanced life. Air would chill as it flowed around the terminator and into darkness. As it cooled to 0 °C the water vapor would snow out. As temperatures continued to fall to –88 °C, carbon dioxide would follow. This would leave a dry atmosphere, bereft of the carbon dioxide and water needed for photosynthesis. However, the process is not so dramatic that the temperatures fall low enough to rain out the nitrogen and oxygen. An atmosphere remains, but the planet becomes uninhabitable to all but microbial life. You could imagine oceans sublimating into the dry air, leaving one side of the planet water-free while the other was encased in an enormous ice cap.

Nightmarish though this scenario sounds it overlooks the ability of the atmosphere to transport heat. On Earth we do not fear the coming of winter in the north. There is no suggestion that as the Arctic plunges headlong into darkness that the planet will soon thereafter be in danger. We count on a vigorously stirred pot of gases to redistribute heat from the tropics to the poles. Moreover, in Europe, winters are commonly mild, as the movement of warm tropical waters offshore ensures that the air blowing off the Atlantic is relatively benign. Given a suitably dense atmosphere – and a circulating ocean – why should we expect the enveloping canopy of gases to falter on a red dwarf world?

It appears that as long as the air is dense enough, sufficient heat can be transported from warm to cool areas through convection and the horizontal flow of air – advection. Thus tidally locked worlds should be able to retain a warm atmosphere, assuming enough radiation is received. Consequently, the threat of the atmosphere becoming so chilled on the dark side that it snowed out is avoided. Therefore, it seems that the majority of worlds within the habitable zones of their parent red dwarf stars will avoid this fate, as long as they have at least a modestly dense atmosphere. Nothing extreme is required.

Liquids move in response to convection heating, and the natural tendency of an atmosphere will be to try and even out any differences in temperature between the sunlit and dark sides. The planetary surface at the SSP may roast at 50 °C, depending on the thickness of the atmosphere. However, vigorous convection in the air above the surface will cause it to move upwards and outwards towards the cooler hemisphere. Meanwhile cold air from the sunless side will flow across the terminator, cooling the day-lit side and allowing warm air to populate the dark hemisphere. As long as the atmosphere is thick enough, air movement should spread sufficient warmth across the bulk of the surface to allow life to flourish.

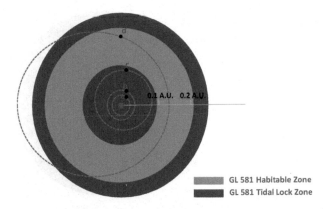

GL 581 Habitable Zone
GL 581 Tidal Lock Zone

Fig. 6.1 The four-planet solution for the Gliese 581 planetary system, based on the original HARPS data. Planets *b*, *c* and *e* are all tidally locked to their parent star, while *d* may be tidally locked or experience very long days lasting almost a full planetary year. A six-planet solution to earlier orbital data has Planet *g* fully within the habitable zone and tidally locked while *d* occupies a circular orbit further out, but still within the tidal-lock radius

Likewise, the models of Heath and co-workers show that even a modest atmosphere, with the density equivalent to that on Earth, would not host unreasonable conditions (Figs. 6.1 and 6.2). The SSP will experience temperatures similar to Death Valley on a hot summer's day, if we bequeath our planet with an atmospheric density one tenth that of Earth. Turn up the atmospheric density to that found on Titan and the bulk of the sunlit hemisphere resides under a tropical 35 °C or so. Even at the terminator temperatures are close to 20 °C (Fig. 6.3).

You might then presuppose that these red dwarf worlds would suffer brutal winds as the planet struggled to redistribute its warmth. Apparently this is not so. Modeling, again, suggests that modest wind speeds of several meters per second are more than sufficient to deliver warmth from sunlit to shade. To put that in context a *good* autumn gale in Scotland has wind speeds in excess of 20 m per second, or roughly 75 miles per hour. What is even more interesting is that for a planet with an atmospheric density similar to Titan, modestly strong airflow between the hot and cold hemispheres ensures that even the permanently dark side never drops below–50 °C. Such temperatures are certainly not pleasant but far from unheard of in Alaska, Canada, Russia or Antarctica.

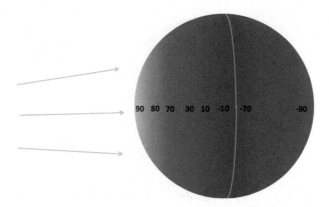

Fig. 6.2 Probable surface temperatures on a planet with an atmospheric pressure of 0.1 bar, one tenth that at the surface of Earth. Heat transfer is relatively inefficient, and the tidally locked sunlit side bakes at nearly the boiling point of water, while the permanently dark side has temperatures more than 160° cooler. *Arrows* indicate direction of irradiation while colors represent temperature, with *blue* being coldest. The numbers illustrate the approximate temperature in degrees Celsius. The *blue line* represents the terminator – the imaginary line on the planet's surface separating the lit and dark hemispheres. It is not the planetary equator

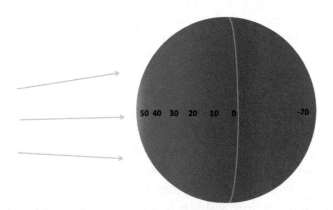

Fig. 6.3 Temperatures on a planet modeled with an atmospheric pressure equivalent to Earth. The thicker atmosphere is far more efficient at transporting heat from the sunlit to the dark hemisphere. The maximum temperature on the sunlit side is considerably lower than in the model, with the thinner atmosphere and similarly, the minimum temperature is not much less than found in an Antarctic winter

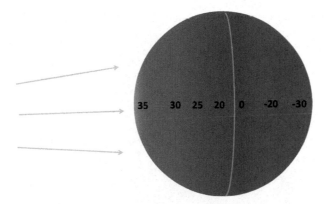

Fɪɢ. **6.4** The final model of Martin Heath of a planet with a 1.5 bar atmosphere, or a pressure 50 % greater than Earth – equivalent to that found on Titan. Temperatures equivalent to tropical Earth dominate the sunlit side, while the dark side has temperatures approximating a cold North American winter. This figure omits one key additional feature. Much of the dark hemisphere hovers around zero Celsius because of efficient heat transfer by winds. This effect is likely to be a regional rather than global effect. This, in turn, is likely to be modified further by topography

Indeed much of this air movement can be accommodated at high altitudes rather than at the surface. Therefore, don't expect extreme storms on tidally locked planets that orbit in their stellar habitable zone (Fig. 6.4).

The Coriolis Effect

All habitable crimson worlds will be tidally locked to their parent star, with a period of rotation of 20–70 Earth-days, depending on the mass of the parent star. What are the consequences of this slow rotation? On Earth – and indeed on every planet in the Solar System that houses an atmosphere – winds are affected by the rotation of the planet. This somewhat counterintuitive response is called the Coriolis effect or, sometimes erroneously, the Coriolis force.

The Coriolis effect is the solution to the phenomenon observed by the German Long Max gunners of World War I, but what exactly is it? The Coriolis effect is the result of a property known as inertia. Inertia can be thought of as the propensity of

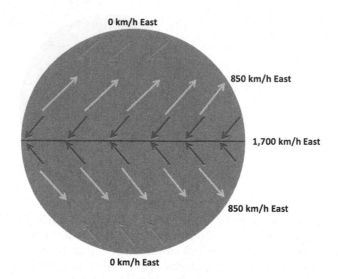

Fig. 6.5 The effect of a spinning Earth on the movement of air to and from the equator. Air moving away polewards is deflected eastward, while air moving towards the equator moves to the west. Numbers indicate the rate of rotation of the surface around the polar axis

moving objects to keep moving in the same direction and speed or, if at rest, to remain so.

Imagine you are standing on the surface of Earth at the equator. Earth rotates on its axis from west to east at roughly 1,700 km per hour – a rather speedy rate. However, at the latitude of 50° north, or roughly the U. S.-Canadian border, the planet's surface has less far to go to complete a single rotation around the pole, so it is effectively moving at less than half the velocity of a point on the surface at the equator. Standing at the North or South Pole you are simply rotating around your middle at a very leisurely pace indeed and effectively you are not rotating around the pole at all (Fig. 6.5).

Now, consider a packet of air moving away from the equator towards the North Pole. As it moves north it is moving faster from west to east than the ground underneath, so it appears to curve more and more to the east as it goes north. Eventually, the air packet appears to cease moving north altogether and moves purely to the east.

On Earth, air rises above the equatorial regions because it is heated strongly from below by the ground. The air rising upwards eventually reaches a lid of increasingly warm air, the tropopause, and is deflected north and south. Instead of making a complete trip

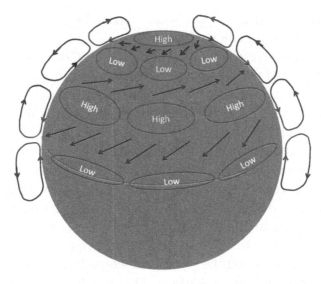

FIG. 6.6 The general pattern of wind flow and pressure on Earth in the northern hemisphere. Winds are organized by both temperature differences between the strongly heated equator and the cold pole, and by the rapid rotation of our planet. Geographical features such as oceans and mountains, strongly affect this generic pattern, as does the tilt of Earth throughout the seasons

to the poles it moves steadily north, curving more and more towards the east, in the northern hemisphere. This is because of the Coriolis effect. At around 30° north (or the equivalent latitude south) it also becomes so cold and dense that it falls back to the surface, arriving as a dry hot airflow over the horse latitudes. This region encompasses the desert regions of our planet. From here, the air either returns south in the northern hemisphere to the equatorial belt or flows northwards towards the mid-latitudes. The air flowing back to the equator is now moving slower than the surface over which it is traveling and, therefore, curves to the west. This gives rise to the northeast trade winds much favored by European mariners en route to the New World. To the south of the equator a mirroring flow of air moves northwest towards the equator (Fig. 6.6).

It is the deflection of air (or indeed artillery shells) due to differences in the relative movement of the surface that is experienced as the Coriolis effect. German artillery missed Paris because they were firing from the north, and, relative to Paris, they were moving east slower than the French capital. The launched shells veered off target to the west, carried away by the difference in the movement of the surface.

Another interesting consequence of the Coriolis effect afflicts those thinking of traveling to Australia from Europe or the United States via a large hole they've dug through the center of the planet. In the stereotypical experiment the person jumps down the hole and accelerates towards the center of Earth. Since energy is conserved, they make it all the way to Australia, arriving at a speed leisurely enough to allow them to step out into the Outback. However, the reality would be rather more grizzly. Thanks to the Coriolis effect the person would steadily migrate towards the wall of the tube as he or she descended toward the center of the planet. Indeed, the effect would be so pronounced that they would smash into the wall of their tube – at some considerable speed – only a kilometer or so below the surface. This would be a rather unpleasant end to a trans-planetary journey.

On top of the straightforward Coriolis effect is the simple effect of motion relative to the surface. Air moving high up, parallel to the surface, experiences a "centrifugal force." This is another effect of inertia, and once again the term "force" is erroneously applied. Centrifugal force increases with the speed of movement.

To get an idea of what this force is recall your experience in a car as it travels around a bend in the road. You feel a force pulling you towards the outside of the bend as the car veers. This isn't really a force pushing you at all – it is your inertia trying to carry you forward in the same direction you were moving in originally. This acts against the turning motion of the car. Likewise, in the atmosphere, the centrifugal force wants the air to move continuously in a straight line, while gravity wants to pull it vertically towards the surface. The centrifugal force acts at right angles to the axis of Earth's rotation and thus has a sideways pull. The net effect is that air moving towards the east is moving faster than the surface over which it flows and thus curves southward towards the equator. Air moving towards the west has less centrifugal force than the surface and is deflected towards the pole.

If the Coriolis effect appears counterintuitive, it can easily be demonstrated on a carousel. Take a ball and roll it across the surface of the carousel when it is stationary, and it appears to move, as expected, in a straight line. But repeat this while the carousel is spinning and the ball appears to curve to one side as it approaches the center of the ride and back again as it moves to the opposing side. This is an effect of the carousel surface moving at a different speed relative to the ball.

On any spinning world, air is deflected by this force. On Earth, as elsewhere, air moving from the equator towards the poles has to move to eastward. This generates the strong prevailing westerly winds of mid-latitudes. From the poles cold air drifts southwards towards warmer climes, driven by convection of the air further south. As this cold air is moving over a faster moving surface, the air moves westward relative to it. On Earth, or indeed any large spinning body, the air flow from warm to cold is thus broken up into a number of westward or eastward moving air belts.

The faster the planet spins the less easy it is for air to move from warm to cold, and the air flow contains more and more belts. Look at Jupiter and this effect is very evident. The air moving from warm to cold travels a relatively short distance north or south before it is diverted by the Coriolis effect. Further belts of air develop alongside these until air movement is able to effectively transport heat from warmer to cooler climes.

As the Coriolis effect is dependent on the speed of rotation of the planet, motion within the atmosphere of a habitable world will be a very different effect to that on Earth. Earth rotates once roughly every 24 h. On a planet orbiting a red dwarf, with one third the mass of the Sun, habitability means that one orbit will last roughly 30–50 Earth days. This figure drops to 6 days for a planet orbiting a red dwarf with a mass one tenth that of the Sun. For the more massive 0.3 solar mass star the tidally locked planet must rotate once on its axis for every orbit; thus each day will be 30–50 times longer than here on Earth. This makes the Coriolis effect much weaker – roughly 15–25 times less effective than on Earth. Air is, therefore, deflected much less severely by it than on Earth. As a consequence, air moving from areas of high to low pressure (and vice versa at high altitudes) will move more strongly across lines of pressure than happens on Earth.

Does this mean that winds will be sluggish on a planet orbiting a red dwarf, compared to those found on Earth? Or does it mean that atmospheric circulation will be profoundly different? An interesting comparison can be made with Venus. Our sister world also rotates slowly – once per 243 Earth days – or a little over 21 days longer than the time it takes to orbit the Sun. The direction of rotation is also opposite that of its orbital motion. Consequently, the Sun rises in the west and sets in the east, giving Venus some interesting analogies with those worlds orbiting red dwarfs. At the surface winds are sluggish, but high up powerful winds blow in the opposite direction to the planet's surface. This "super-rotation" may in fact be common within the atmospheres of planets orbiting crimson suns. Indeed it appears to be this way in some of the models that are described below. Overall, despite a slow rotation, the atmosphere of Venus is effective at transporting energy. In large part this is due to the high density of the gases. Thus, when considering the circulation of the atmosphere, the density and depth of it will also be important. This will be all the more true for crimson worlds, where there are strong differences in heating across the terminator.

Atmospheric Modeling

Work by Manoj Joshi (University of East Anglia) and, subsequently, Tim Merlis (Princeton University) has examined the flow of air on both aquaplanets – water-enshrouded worlds – and those with drier surfaces. In the models of both Joshi and Merlis, movement of air from warmth to darkness is accommodated in a fairly complex pattern consisting of several different belts. Figure 6.7 attempts to simplify the findings of this work, showing where areas of different air flow occur on a planet that wasn't spinning and therefore lacked a Coriolis effect.

However, on any realistic planet there is rotation, so the outcome of this model is unlikely to be directly applicable. As we already know, although the Coriolis effect is weaker on planets orbiting red dwarf stars it is still present. For a planet orbiting a 0.08–0.1 solar mass star, its day would be approximately 144 h in length, and the Coriolis effect will be one third as strong as on Earth. Thus, ironically, a planet orbiting the lowest mass red dwarfs will have days more closely approximating that of Earth and the winds blowing west to east will be correspondingly stronger. Planets orbiting more massive red dwarfs will experience a weaker Coriolis effect as their orbital periods – the time taken to orbit their stars – are longer.

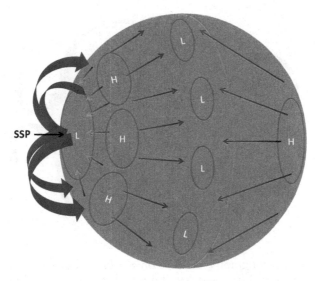

FIG. **6.7** A possible pattern for wind-flow on a planet that does not rotate. Air is heated and rises strongly on the star-lit side (strongest under the SSP) and flows outwards at height. A very simple airflow would involve gases flowing at height from the warmest to coldest regions and back again at the surface – as is seen on Titan. However, with sufficient spin or temperature contrast, high above the surface the air becomes too cold and dense to flow all the way around the planet and much of it descends short of the terminator. Winds flow nearly directly from areas of high to low pressure with no Coriolis Effect. However, even tidally-locked planets do rotate so this pattern would be unrealistic

Even a weak Coriolis effect will modify the simple motion of gases from the dark side towards the SSP, bending them to the east and west, depending on their direction relative to the equator. But that's not all. The additional and quite unusual feature is the flow of air from light to dark sides along the equator, illustrated in Fig. 6.8.

Broadly speaking a planet orbiting a very low mass red dwarf will have an odd pattern of wind flow somewhere between that of the Earth and that of a non-rotating word. Hot air would rise above an elongated region focussed on a point on its equator. This would generate a broadly circular area of low pressure. Ascending hot air would then flow northeastwards and southeastwards at height, while it is replaced by cooler air flowing from northeast and south east towards the SSP. The air moving polewards at height is deflected to some extent by the weak Coriolis Effect, but largely continues unhindered towards the dark hemisphere. Modelling work suggests that the air flow eastwards is weak overall and may be broken into

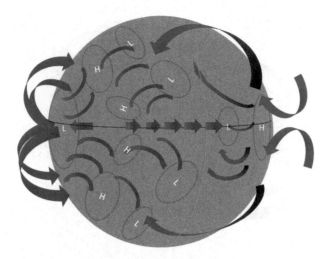

Fig. 6.8 Without topography, airflow on a tidally-locked planet could look something like this. The majority of air still flows across the surface from the dark, cold side to the warm, lit side. Air rises over the SSP at or near the planet's equator (*black line*) forming an area of low pressure. Some of the air cools and descends some distance away forming a belt of high pressure, while the remainder flows across the terminator. With a weak Coriolis Effect airflow is deflected east if it flows towards the equator and west if it moves away from the equator towards the poles. Along the equator a weak current of air flows from the lit to dark sides at height, while over the poles, cold air returns towards the SSP at the surface. The terminator, demarcating the sunlit and dark hemispheres is shown in *blue*

a number of modest cells. Over the equatorial regions, near the top of the troposphere – the lowest and most dynamic portion of the atmosphere – there should exist a westerly airflow away from the SSP. Air will flow into this belt from the surrounding atmosphere in a pale imitation of the converging equatorial air masses on Earth. The majority of cold air flow towards the SSP transits Polar regions at the surface, with much weaker, or reverse, flow nearer the equator. This generates two broad cells, one per hemisphere, stretching from the sun-baked SSP to the dark hemisphere.

Planets with more sluggish rotation appear to have a similar pattern, but with air flowing from cold to the warm, sunlit sides across the poles. A belt of warm, westerly winds blows along the equator towards the dark, cold hemisphere. Either way there is sufficient spin in these planets to generate westerly and easterly winds, rather than the expected simple surface dark, cold to warm, lit sides and vice versa at heights.

Why is there a westerly air flow over the equator in the models of slow rotating planets? This counterintuitive flow is produced in response to the strong difference in the amount of heating across each hemisphere of the tidally locked planet. On tidally locked worlds strong longitudinal variation in heating generates a wave within the atmosphere called a standing Rossby wave. Standing waves are popular objects for demonstration with the toy Slinky, ripple tanks and other pieces of apparatus dragged out for physics lessons in high schools. Although the material the wave is built from may move across the globe, the positions of the wave's crests and troughs remains fixed. They stand still, hence the name. In the case of standing Rossby waves, long wavelength atmospheric waves remain locked geographically, despite the air moving along and through them.

Adam Showman (University of Arizona) and Lorenzo Polvani (Columbia University, New York) illustrated this effect in models of the atmospheres of tidally locked hot Jupiters and smaller terrestrial worlds. Strong heating on the sunlit hemisphere generated standing waves in the upper atmosphere, and these had the effect of driving air eastward over equatorial regions of these tidally locked worlds.

A weak westerly jet stream is known on Earth over the equatorial regions, but on a tidally locked world, this feature is likely strong, transferring atmospheric warmth at a significant rate. Such westerly airflow, counter to the rotation of the planet, is known as super-rotation and is likely a common feature of the atmospheres of tidally locked planets. Subsequent to modeling, infrared observations of some hot Jupiter worlds showed a clear displacement of the highest temperatures from the SSP by several tens of degrees longitudinally. For example this displacement was beautifully illustrated in observations of the hot Jupiter HD 189733b by the Hubble Space Telescope. Thus at low speeds of rotation, planets experience an equatorial westerly wind, but this reverses as the speed of planetary rotation increases and the Coriolis effect strengthens.

On tidally locked planets, such as all habitable planets orbiting red dwarf stars, much of the westerly airflow is likely accommodated at high altitudes, unless the atmosphere is thin. However, at the surface gentle westerly breezes at the equator or over more poleward latitudes may permeate the terminator, usefully bringing warm air from the sunlit to the dark hemisphere. Thus temperatures in the permanent, Stygian gloom may be moderated.[1]

The Effect of Topography

On Earth the most significant annual atmospheric phenomenon is the Asian monsoon. In the winter the continental surface cools dramatically, and cold air flows outwards towards the warmer Indian Ocean. The tilt of Earth and its motion around our star causes the Sun to shine strongly down on the southern hemisphere during the northern winter. However, as Earth moves around in its orbit and the northern summer approaches, the Sun moves higher in the sky. As the spring progresses the Sun shines more and more strongly on the Asian continent.

As the Sun projects more strongly at a steeper angle and the length of day increases, the land rapidly warms up. Air pressure steadily falls over India as the hot land stirs the air above into powerful convection currents. In June a "front" of cooler, humid air advances northward towards the Himalayas, reaching the north of India and southern China in early July. With the flow of moist air comes the life-giving summer rains – the summer monsoon has begun. Simultaneously, an easterly jet stream develops over northern India and Tibet, steering some summer storms along with it.

Although there are no seasons on tidally locked planets, variations in the distribution of ocean and continent will strongly modify airflow and hence the types of weather experienced at each location. Continental regions will always heat more strongly than neighboring oceans. In part this is because the land has a lower specific heat capacity than the oceans – essentially meaning that less energy is needed to warm the land than the sea. But also the land is fixed and unmoving, except on geological timescales. The oceans, by contrast, are fluid and circulate, allowing them to redistribute the energy they receive from their star.

---

[1]Dr. Jun Yang (University of Chicago) has modelled atmospheres on tidally-locked planets and these have a cloudy sun-lit side and largely cloud-free dark side. The cloud serves to lower temperatures on the lit hemisphere.

Thus, in the warm, illuminated hemisphere any land will heat up more strongly than neighboring oceans. Hence air will flow from cooler ocean areas towards any land exposed to the star. Similarly, in the colder dark hemisphere any land areas will be the coldest, as its low specific heat capacity allows it to shed energy effectively. Air will cool over the land and flow outwards towards the oceans. This movement will be deflected to varying extents by the weak Coriolis effect on these planets. Moreover, circulation of cold and warm water within the oceans will ensure that the oceanic areas are warmest on this hemisphere. These differences in temperature will generate regional wind patterns, with air flowing from colder surfaces towards warmer ones. This will obviously complicate the grossly simplified pattern of circulation shown earlier in Figs. 6.7 and 6.8.

Most importantly currents of warm water moving to the night side across the terminator will bring much higher temperatures, while cold water moving towards the sunlit side will lower temperatures there. Despite an apparent uniformity in terms of heating of tidally locked planets longitudinal and latitudinal variations in temperature will still exist as a result of circulation within the planet's oceans and atmosphere and variations in land distribution.

Gliesian meteorologists and cartographers, should they exist, might have a three-point mapping system: a system of latitude extending north and south from the equator; longitude around the axis of rotation, as on Earth, but perhaps zeroed in on the point where the SSP aligns with the equator; and a circular grid extending away from the SSP that would allow Gliesian sailors to navigate with their Sun.

On land-free aquaplanets, surface heating over the most strongly lit hemisphere will generate strong, moist convection and rainfall near the SSP, while air flows more or less evenly across the surface from areas of high to low atmospheric pressure. Without land, air flow has few interruptions, and the corresponding meteorology is simpler. However, the role of the seemingly contrary airflow from warm to cold hemispheres shown in Fig. 6.8 will still modify climate and likely drive a pattern of circulation in the ocean that helps transport heat to the dark hemisphere.

The exact pattern of air flow will also reflect the pattern of topography where there is land, the distribution of any land and water, the period of rotation of the planet, and the amount of heating the planet receives from its parent star. More obviously, any mountains will intercept sunlight so that the sunlit slopes of these heat up strongly, generating their own belts of low pressure, much like the Himalayas during a northern terrestrial summer.

Mountains will also shade the regions behind them, causing significant cooling. Finally, mountainous regions spanning lines of latitude will disrupt the easterly or westerly airflow from sunlit to dark hemispheres, or vice versa. This may have the effect of causing dramatic cooling on the dark hemisphere, with global consequences for atmospheric temperature and circulation. Moreover, mountains will intercept any moisture in the air, generating both heavy precipitation on their windward side and a rainless shadow in their lee. Therefore, meteorology on a planet with land and water is unlikely to be as straightforward as it might appear at first glance. An alien forecaster could well have his or her work cut out predicting the daily weather.

Conclusions

Climate modeling on Earth is frequently a contentious affair, with even the seemingly simple assertion that increasing the concentration of the greenhouse gas, carbon dioxide, increases temperature. This is with a substantial body of research and computational power backing it. Modeling the atmospheres of planets orbiting red dwarfs is at a more primitive stage. Models have been run periodically over the last two decades, often with similar results. The location of regions of varying temperature has led to modeling of atmospheric motion on aquaplanets and more terrestrial-like worlds with land and sea.

However, models have a fairly low resolution and don't take into account variations in topography at present. Once we have spied on a few of these habitable worlds with instruments that will pale Hubble and Kepler into the shade, we may well be able to apply these models to real alien environments and test their efficacy. Moreover, when we have data from these alien worlds at our disposal we can apply information from them to our world. Perhaps then we will convince a few climate skeptics that changing carbon dioxide concentration really does have an effect on global temperature and that this is in line with current terrestrial observations.

7. The Development and Sustenance of Life

Introduction

To the human eye life seems a delicate, even an ephemeral, thing. After all, we occupy such a short, whimsical fraction of astrophysical time. We are so much less than the candy sprinkles on the universal cake. Yet our common perception of living organisms belies a deeper understanding of life itself. Although organisms expire with the same apparent ease that brings them into being, life itself is utterly robust. Even simple investigation shows this to be true. The dinosaurs were wiped off the planet 65 million years ago by the impact of a 10 km-wide asteroid. Yet this observation disguises the rapid resurgence of life within a 1,000 years or so of the catastrophe. Fossilized fern spores show that almost as soon as life was able to return, it did. Plainly, many organisms had no problem with the annihilation of the dominant species. Indeed, they prospered as a result.

More catastrophically, the Permian mass extinction, 251 million years ago, followed a super-volcanic outburst in Siberia and global oceanic anoxia – the loss of dissolved oxygen. A full 90 % of species were obliterated. Yet within a few million years the stage was set for the rise of the dinosaurs, mammals and later the birds. To use the eponymous phrase, life finds a way. Life is utterly robust.

However, can we apply our common experience of life on Earth to other, distant worlds? Do the same underlying principles apply, and if so, what predictions can we make about our alien cousins, if indeed we are related at all?

D.S. Stevenson, *Under a Crimson Sun: Prospects for Life in a Red Dwarf System*, Astronomers' Universe, DOI 10.1007/978-1-4614-8133-1_7, © Springer Science+Business Media New York 2013

Our Alien World: The Hot, Deep Biosphere

In the 1980s Thomas Gold made what seemed at the time a prepos-
terous assertion that living microorganisms would dominate the
deep strata of Earth. His "hot biosphere" was hardly lauded as the
work of genius at the time of its publication. Yet the discovery of
microorganisms living in deep, dark, volcanic ocean vents – more
bizarrely, thousands of meters down in the Columbia River basalts
or the gold mines of South Africa – illustrates convincingly that if
life has a chance, it takes it. Organisms are adept at finding niches
in which they can survive.

Sunlight is not essential for life, nor is organic material from
which to grow. Microbiological life is robust, highly adaptive and
ingenious at solving problems. Life, invariably, finds a way.

Our experience simply tells us that the only common denom-
inators for the development of life are water and the presence of
an energy source. The maxim: "follow the water" drives current
thinking in most astrobiological circles.

The Concept of the Stellar Climatic Habitable Zone

Astronomers, somewhat belatedly, took on the idea of biological
robustness. Until fairly recently, the idea of the "stellar climatic
habitable zone" (SCHZ) dominated thinking about where extraterrestrial life would be found elsewhere in the universe. A planet,
were it to host life, would need to be positioned at just the right
distance from its parent star, ensuring it received the correct
amount of warmth needed to maintain liquid water on its surface.
At reasonable atmospheric pressures water is liquid between the
temperatures of 0 and 100 °C. Since water (at least on Earth) is
an apparent pre-requisite for life, find a planet at a distance from
its star with temperatures in this range, and there will at least be
the opportunity for life. The concept of a habitable zone existing
around a star is therefore – at least superficially – a sound one.

However, the discovery of exotic microbial life in apparently
outlandish and extreme environments on Earth has thrown this
assumption into question.

The location of the stellar climatic habitable zone around any star is easy to determine. If you know how much energy a star radiates – again, something that is relatively simple to measure – then using some simple mathematics you can work out the average temperature of a planet at a given distance from its host. This math is known as Stefan's law and is something high school students routinely do. Yes, there are some variables, such as the type of world, its reflectivity, the presence of clouds, etc., that need to be considered, but the math remains fairly straightforward. In our Solar System Earth lies fairly deeply within the Sun's climatic habitable zone, with Mars lying near its outer edge. Venus lies on the hot side of this zone and is thus boiled dry.

However, things are not so simple. Even a cursory examination of Mars reveals that it once had a surface drenched in water. This presents a small problem, as early on, when Mars was wet, the Sun was cooler and hence dimmer. Mars should have been frozen, lying outside the outer edge of its habitable zone. Therefore, even at this stage, the idea of a stellar habitable zone requires some tweaking.

Despite these exceptions, planets within their stellar habitable zone should, in principle, be the first places to look for life. If we exclude large gaseous worlds, the remaining terrestrial (rocky) bodies would be expected to host the conditions needed for liquid water and life. In 2007 the star Gliese 581, the central star of this book, was found to host four and possibly six worlds. Two of them, Gliese 581d and better still Gliese 581g, appeared to lie within the stellar habitable zone. Although considerable doubt has been attached to the existence of Gliese 581g, Planet d remains a candidate as a habitable world, at least in terms of its potential surface temperature. Elsewhere, a number of other worlds that have been identified by the Kepler mission appear to lie within the stellar habitable zones of their stars. In principle these could host life, if they have sufficient liquid water on their surfaces.

What Is Life?

Let's accept the premise that life requires liquid water. How then does simple chemistry evolve into a series of complex biological machines that can talk, walk and think? In order to understand this we must first turn back the clock and look at the sorts of environments in which biological chemistry may have arisen.

Imagine a world bathed in liquid water. Its atmosphere is mostly carbon dioxide and nitrogen, with a few percent of other gases such as water vapor and argon. Ultraviolet light permeates rolling clouds of moisture, while the surface cracks under a frequent bombardment by comets and asteroids. Over 95 % of the land is submerged in salt water, with only small volcanic islands projecting haphazardly above this sterile sea. Lightning crackles from frequent storms, many triggered by volcanic outbursts. In the depths the thin crust is fractured extensively by impacts and by the continuous effect of convection within the hot, deep interior. Magma erupts deep under the ocean surface, building a dense, thick, basalt and komatite crust. In places this thick, dense solid scum fractures and descends under its own weight, back into the hot mantle, generating even more volcanism.

Welcome to the infant Earth. It is a forbidding place, bereft of the kind of life-giving properties we take for granted. Yet around undersea volcanic vents, and in those same boiling volcanic plumes, gases are starting to come together in ways that will eventually allow life to develop.

At undersea vents, reactions between ubiquitous iron, sulfur, water and carbon dioxide are generating methane and other, more complex organic materials. Within the atmosphere, the ubiquitous plumes of volcanic gases are drenched in lightning and ultraviolet light. Complex reactions are brewing reactive compounds of nitrogen and oxygen. Rain corrals these chemicals and dumps them into the oceans, blending them with the soup that has been brewed by undersea volcanism. Into this Earthly stew comets and asteroid impacts deliver further complex organic molecules that have been synthesized cosmologically. By the time Earth is a few 100 million years old, the oceans are awash with complex organic chemicals, and there is the suggestion of life. But what, exactly, *is* life?

Superficially, defining what constitutes a living organism would seem trivial. We all know from casual observation that mice are alive, sequoias are alive, but lumps of granite are not. Look more closely, and the waters get somewhat muddier. In Britain 11 years olds are taught about MRS GREN. Each letter defines a characteristic of a living thing: M for movement; R for reproduction;

S for stimulation; G for growth; R (number 2) for respiration; E for excretion; and N for nutrition. But look more closely at each and the problems appear.

For example, the growth of a crystal is the repetition of structure – reproduction. So, when that large body of granite solidified and crystals began to grow, did the granite come to life? Probably not. Similarly, when crystals are seeded into a saturated solution of some salt and they get bigger, is this growth? So, we can remove these two as defining features of life.

Look at the other features embodied in MRS GREN, such as movement, and things get worse. Rocks tumbling down hills under gravity are definitely not alive, while trees don't much move. Plants will, however, respond to environmental stimuli, such as gravity and light, and move or bend in response. Does this mean trees are dead, or is the tumbling rock – also responding to gravity – alive?

On a larger scale an asteroid in orbit around the Sun clearly moves in response to gravity and also to sunlight through the Yanofsky effect. Is it alive? Nutrition? Again, you could think of the growth of crystals from a solution as a form of nutrition. Within some living organisms there are protein structures called microtubules and others called microfilaments. Both of these grow by "fixing" building blocks from the cellular broth. However, these are clearly not alive in their own right. Excretion is more tricky, but if we include in excretion the process of loss of volatile substances, gases, or other small molecules, bacteria can be accommodated, but so can many chemical reactions such as combustion. Is a piece of burning coal alive?

Finally, that leaves "stimulation" – a response to some external factor such as heat, light or sound. We've already seen that asteroids respond to sunlight, so we are already on shaky ground.

What's left? There is one characteristic, shared by all living organisms, that appears unique to them: evolution. All living organisms, including viruses, evolve. Evolution, through natural selection, is merely the change in the structure and function of an organism over time in response to natural factors, including mutations. Only living things evolve. Evolution involves changes to the coding sequence of the genes that instruct the organism. This, in turn, changes how the organism works. Neither individual crystals nor entire asteroids evolve. An asteroid may change its structure in response to a violent collision with another, but it will still be made up of the same material with which it started out.

Mutations are for all intents and purposes permanent changes in the coding of an organism – changes that are then selected by different environmental factors. Therefore, in the context of this book, we will think of living organisms as structures that can evolve and change over time. Take them apart and look more closely, and you will see that a living organism is an evolving bag of chemical reactions. This may not be entirely glamorous, but it is fairly close to reality.

How Does Life Arise?

With an eye on the characteristics of life let's then return to our infant planet and its oceans awash with life's building blocks. The science of this scenario is robust. There is extensive evidence that comets and asteroids contain abundant organic material – the kind that can produce living organisms. Moreover, research indicated that simple reactions involving dissolved gases, water and the iron- and magnesium-rich mineral olivine, allow the synthesis of methane. In turn, under the bombardment of ultraviolet light, methane tends to react with itself, water and nitrogen to form more complex amino acids. This sort of complex synthesis is evident in the atmosphere of the much colder Titan as well as in the deep, cold molecular clouds from which stars and planets are born. There is no reason to doubt that it could have happened on a much more clement Earth (Fig. 7.1).

There is at least circumstantial evidence that by 3.8 billion years ago there was some form of life on Earth – this despite the fairly hellish conditions. The evidence is not fossil-based but rather isotopic. Living organisms work through the actions of enzymes – proteins or molecules of ribonucleic acid. These catalyze chemical reactions within the cell that keep it alive. Such molecules grab hold of certain precursors, known as substrates, and selectively combine them to make other compounds. Enzymes, however, are a little choosy about substrates. For example, carbon comes in three flavors. The majority of carbon has a mass of 12 units

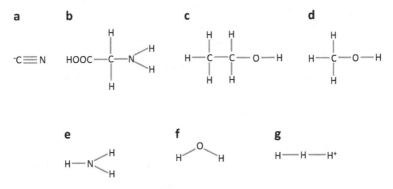

FIG. 7.1 Common molecules found in interstellar space that are important for the generation or sustenance of life. (a), cyanide; (b), the simplest amino acid, glycine; (c), ethanol; (d), methanol; (e), ammonia; (f), water; and (g), a reactive form of hydrogen produced by the action of ultraviolet light in nebulae. This reactive molecule appears necessary to produce the vast treasure trove of molecules found in interstellar space

(carbon-12). However, a much smaller proportion is somewhat heavier – carbon-13. The process of photosynthesis preferentially absorbs carbon dioxide containing the lightest and most abundant isotope of carbon. In general life on Earth tends to prefer this isotope, thus imparting a chemical bias on the carbon preserved in fossilized material. The Isua formation in Greenland has such a bias, suggesting life existed there and was the source of this chemical shift. However, these rocks have had a rough time since they were initially laid down near a subduction zone. Metamorphism of the material appears extensive, and it may well be that the life-like alteration in carbon isotopes is a mirage caused by distinctly non-biological processes. If so, life arose much later, perhaps as much as 300 million years, after the formation of Isua, or 1 billion years after the planet was born. At this stage we begin to see what look like fossilized bacteria and some more chemistry suggestive of life.

How does life emerge? Is it an inevitable process in the evolution of a planet? And when it does emerge, will it follow the same path?

These are key questions for Earth, never mind elsewhere. As a biologist this author might want to answer the second question first. The development of life does seem inevitable – at least on every planet that has conditions conducive to its survival. As for the third question, the answer to this is undoubtedly negative, but

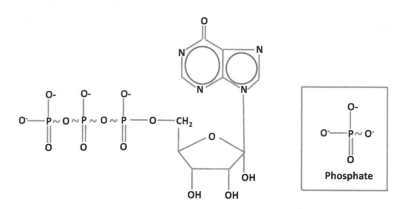

Fig. 7.2 The structure of the cellular energy molecule adenosine triphosphate, or ATP. Energy is stored in the chemical bonds that link the phosphate groups together (inset). These are shown by *wiggles* rather than *straight lines*. On Earth, ATP stores chemical energy from glucose (respiration) and from sunlight (photosynthesis) in all known organisms

there are likely common themes running through living organisms on every planet. This will be explained in more detail later.

Pursuing the second question, life in simplest terms could be reduced to the term "applied chemistry." Even processes we regard as complex, such as contemplating how many doughnuts to have with your coffee, comes down to a series of orchestrated chemical reactions. Thinking requires energy in the form of a chemical called adenosine triphosphate, ATP for short (Fig. 7.2). ATP is the currency of all life on Earth. ATP is the biological energy store synthesized by certain chemical reactions in our cells. One of the most familiar of these orchestrations is the process of respiration. Another, unique to plants and some bacteria, is the equally familiar photosynthesis.

Cells use ATP to power the manufacture of useful substances such as proteins, fats and carbohydrates. ATP is also used to power the movement of cells and the components within them. Thinking is a series of steps involving the movement of charged particles of sodium and potassium across membranes and the linking of nerve cells with chemical messengers. Moving your hand to lift doughnut number three to your mouth requires ATP to power the muscle contraction. These contractions lift your hand and allow you to use your opposable thumb and fingers to grip your cup. Chewing and digestion are simply a series of further chemical reactions.

Living organisms are merely bags in which chemistry happens in a controlled and modest way. Biology allows often highly dangerous chemical reactions to happen safely and in concert with others that wouldn't normally happen in unison. By controlling it, life manipulates energy, turning chemistry into a tool that allows for reproduction as well as growth. The development of life is simply the reorganization of chemistry from the chaotic to the sublime.

With chemistry in mind we can now ask, how did life on Earth begin? The short answer is we don't know, but there are clues. We already know that the chemical materials needed to get life going was probably in place soon after Earth formed. Nothing spectacular or unusual was needed. The stars provided the chemical building blocks in abundance, and these were supplied in profusion during planet formation. Within the molecular clouds in which stars and planets form, we have detected abundant amino acids, alcohols, hydrocarbons and other bits and pieces of useful, pre-biotic (pre-life) organic chemistry needed for the synthesis of life. At the other end of the process we also know a fair bit about how cells have evolved over time. However, there are many complex steps between the pre-biotic chemistry of the interstellar cloud and even the simplest cell (Fig. 7.3).

One of the biggest problems for the origin of life is the development of genetic coding. In all Earthly life forms, viruses included, the genetic information is carried on one of two types of molecule. For bacteria, many viruses and all complex life, genetic information is encoded on a type of molecule called a nucleic acid. For these organisms, the nucleic acid is the eponymous DNA – deoxyribonucleic acid.

Some viruses, such as H.I.V. and influenza, use a slightly different nucleic acid, ribonucleic acid or RNA. In turn the genetic information these molecules hold is "read" primarily by a set of machines built from entirely different kinds of molecules – proteins – but with considerable assistance from RNA. By "reading" we mean the conversion of the language of the genetic code – the information held on the molecule – to another found in proteins. Biologists use the terms "transcription" and "translation" to describe these two stages in the transfer of information from DNA to protein. Proteins are in turn the workhorses of the cell, carrying out all (or most) of the chemical reactions that keep the cell alive.

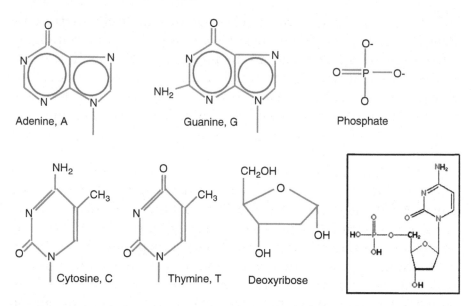

FIG. 7.3 Stylized illustrations of the building blocks of DNA, deoxyribonucleic acid. Each unit, called a nucleotide, is made up of a unit of phosphate, deoxyribose sugar and one of the four, more complex molecules, *A, C, G* or *T*. The structure of one nucleotide is shown in the inset, *bottom right*

    In all Earthly cells that we have found the genetic code consists of two pairs of letters bound in the famous double helix. In DNA these are A-T and G-C pairs: A stands for adenine, T for thymine, G for guanine and C for cytosine. Each set of letter pairs is copied, through the process of transcription, to produce a molecular transcript, messenger RNA. Here, the code is slightly altered, with T being replaced by another, closely related compound denoted by the letter U. This messenger RNA, or mRNA for short, passes to a complex machine called the ribosome, made of RNA and protein. The ribosome then assembles amino acids in the correct order to produce a protein. The trick here is that the ribosome must convert, or translate, the genetic code carried in the mRNA to the new language of amino acids found in proteins. Complex though this all sounds, the processes are fairly well understood, with the vast majority of the key components having had their internal structures probed and mapped using X-rays and electrons.
    Beautiful though the construction of all this machinery is, there exists a small problem at the heart of biology, one that is not yet resolved in its entirety. In all cells and viruses RNA and

DNA require proteins to copy them. Proteins need DNA or RNA to code for them. Moreover, proteins cannot copy themselves, but the machinery needed to copy consists of proteins. Therefore, how do you build a cell that has a genetic store that requires proteins to construct and read it, while simultaneously the proteins that operate it need a functional code to inform their assembly? One cannot come before the other, or can it?

Most biologists believe the solution to this conundrum lies with RNA, ribonucleic acid. Several years ago this author published an article that suggested a mechanism that would bridge part of this gap between proteins and RNA or DNA. The scenario allowed RNA to bridge the gap between the two halves of the cell's machinery. In essence, those little factories that use messenger RNA to direct the manufacture of proteins were the missing link in the evolution of our modern cells. The model envisaged the protein factories as living fossils, imprints of an earlier replication machine.

How might this have been possible? Although RNA is much less effective as a genetic store than DNA, the molecule can still retain sizable amounts of information. Moreover, crucially like protein, RNA can also serve as an enzyme and carry out important biological tasks, including copying itself. These RNA enzymes are known as ribozymes. Most significantly, the key enzyme in the cell that manufactures new proteins, the ribosome, is itself an RNA-based enzyme. The implication is that on the early Earth, RNA or something similar, served as both the store of information and the machinery that used it. In the 1960s and 1970s it was noted that many of the B-vitamins, key components in the metabolism of cells, are attached to parts of RNA molecules. This implies that the key biological processes that generate energy were intimately linked to RNA, rather than protein, in the distant past. There is no other reason to have this set up. The RNA component of the vitamin contributes nothing directly to the metabolic process that the vitamin carries out. The RNA component of the vitamin is apparently a fifth wheel. However, if it was needed in the past to link the functional vitamin to the RNA that orchestrated the reaction, its presence in the modern-day vitamin would make sense.

Although speculative, it seems likely that the chemistry of the early Earth organized itself into some form of primitive molecule that was both genome, the genetic store, and enzyme, the machinery needed to copy it. Later on, further evolution resolved these functions into two distinct camps, each more efficient at carrying out their particular role than the original, single molecule. Whether this molecule was RNA or something even simpler is utterly unclear, and it may well remain so unless we find some molecular fossil hidden somewhere on Earth – or perhaps on another body in the Solar System.

Our understanding of the origin of life is, and may remain, inspired guesswork. The key driver in the division of labor was evolution through natural selection. RNA is quite good as a genetic store and as an enzyme. DNA is very good at storing information, and proteins are very good as enzymes. Although there is no foresight in evolution, once you start fiddling with the RNA so that it can produce proteins, perhaps small ones at first, then the stage is set for evolutionary processes to step in.

Imagine a soccer team consisting of 11 defenders. Defenders can stop goals and they can score them, but they are not there primarily to do these jobs. However, if you allow a bit of natural selection to occur some defenders might spend a bit more time at the front. This puts them in a better position to score goals but a worse one to stop them. You get the idea. In this instance the selection pressure is winning the match. In biological evolution it is reproducing. In the scheme the author suggested in 2002, the ribosome, that machine that manufactures proteins, began its life copying the genetic material, RNA. As the cells became more complex the process was doctored so that proteins were produced alongside the copied RNA molecule. This eventually allowed RNA to become the genetic material and proteins the machinery. Certainly, this theory is speculative, but it links the process of natural selection to function and development of the ribosome.[1]

From Stellar Fluff to Complex Cells

The building blocks of proteins and other components of cells are found ubiquitously in the cosmos, and there are ample opportunities during planet formation to deliver these to a new world

---

[1] Stevenson, D.S. 2002. Co-evolution of the genetic code and ribozyme replication. *Journal of Theoretical Biology* 217(2): 235–53.

either at its inception or soon thereafter. Given the ubiquity of the components that comprise these molecules, it seems probable that similar events to those on Earth can unfold elsewhere in the universe. Proteins, in particular, may well be universal. Their building blocks, amino acids, are common molecular fluff and should be available in most if not all developing planetary systems. There is no reason to believe they should participate exclusively in life processes on Earth.

RNA is made from more complex stuff. Chemicals called pyrimidines (C and U in Fig. 7.2) and purines (A and G in Fig. 7.2) comprise the portions of the molecule that hold the genetic code. These are linked through a simple sugar called ribose and phosphate to one another. Ribose is a simple variant of deoxyribose that is shown in Fig. 7.2. The pyrimidines and purines can be synthesized in interstellar space through the action of ultraviolet light on water and cyanide. Ironically this highly toxic gas can combine with itself to make part of the building blocks of life.

Sugars are more complicated to make, but some simple ones have again been found in meteorites that have fallen to Earth. The implication is, life had many of its raw materials supplied very early on when Earth was in its infancy. Why, then, would conditions around a red dwarf be much different? There is no real suggestion that they would. Although the discs that form the star and any accompanying planets are lower in mass and hence more fragile, the composition of the discs is unlikely to vary much from star to star, even over long stretches of time. This is a result of fairly thorough mixing of stellar gases in galaxies. What we do know is that Gliese 581 is older and somewhat lower in metals – heavy elements – than the Sun, but not by much. Therefore, the planets of Gliese 581 and 667 are likely to have a broadly similar overall composition to the planets orbiting the Sun.

There is one caveat here. Older stars tend to form from material somewhat less rich in iron than younger ones. This is a consequence of the way stars of different masses die. The first generation of stars in the universe appear to have had higher masses than later generations. Without going into the detail of this, it means that when these stars died they released a burst of heavier elements that differed in composition from the slower burning, lower mass stars. High mass stars tend to produce more elements such as oxygen, magnesium and calcium when they explode in so-called Type II supernovae. Lower mass stars release a lot of carbon, nitrogen and oxygen when they expire as planetary nebulae. When some of their numbers explode as Type Ia supernovae, the iron, silicon and sulfur, found ubiquitously in the mantles and crusts of planets, are produced in abundance.

Type II supernovae produce the lion's share of radioactive elements such as thorium and uranium that serve to maintain warmth in the interiors of planets. These elements, with a sizable contribution from some isotopes of potassium, are therefore necessary to maintain an active geological cycle on planets such as Earth and presumably larger super-terrans. Thus the age of a star (and its worlds) is impacted by the material from which it is born. At the moment we haven't analyzed enough of the potentially habitable systems to determine what sorts of elements are present, but it should be taken into consideration when we consider habitability. The life of the biosphere depends on the life of the planetary interior, the geosphere.

So, at this stage, we have good evidence that life can get started with relative ease. At least on Earth, there is isotopic evidence that it appeared close to the end of the period of heavy bombardment. Up to, or possibly peaking at, 3.9 billion years ago, Earth was thoroughly pummeled by comets and asteroids. Some of those would have been big enough to at least temporarily boil the oceans and heat any dry land to high temperatures. Life may have clung on in deeper water that was unaffected by such catastrophes, but it would certainly have had a brutal time of it. Yet, the Isua rocks of Greenland appear to show evidence for life within 50 million years of the end of the bombardment. This is a blink in the cosmic eye. Such a short interval of time conceivably affords all planetary systems, bar those orbiting massive stars, the potential to form living organisms of one sort or another, before their host star expires.

However, we only have Earth as an example at the moment. Although unlikely, it could be a unique world. Applying the Copernican principle, we are average, and other worlds will have similar luck to any other world, given similar initial circumstances. This is something of a leap of faith, but it is not unreasonable, and hopefully in only a few years Kepler, Hubble, or another successor satellite system will confirm the existence of life on other worlds. It is really only at this point you can start pontificating with any certainty on the rarity or abundance of Earth-like habitable worlds. Anything else is premature.

Given the apparent ease at which life began on Earth, we can tentatively conclude that the formation of life on a planet is likely to be highly probable given appropriate circumstances. This is certainly a fairly weak conclusion at present, but not unreasonable.

Finally, let's answer that third question. Will life invariably follow the same path? To answer this one needs to examine the evolutionary history of life on Earth, and in particular chance events that seem to be associated with major episodes of evolutionary change.

If we examine, for example, the rise of mammals, two extreme, chance events are important. These are the Permian mass extinction and the extinction of the dinosaurs. Both appear to have very different underlying causes: extreme volcanism and a cosmic collision, respectively. The Permian mass extinction obliterated many of the ruling reptilian life forms, paving the way for the warm-blooded vertebrates, with the mammals soon after and birds millions of years later. However, for nearly 200 million years the dinosaurs kept the mammals in check, repressing their development, stunting them at shrew-sized scavengers. The elimination of the dinosaurs removed the competition for space and food, allowing the mammals to thrive and move into new environments and evolve, leaving us with the world we see today. This is all down to chance.

We also owe our modern world to evolutionary processes in plants. Modern humans rely on agriculture, and this is intimately associated with the rise of grasses. The grasses represent a cunning adaptation to changes in the amount of carbon dioxide gas in the atmosphere. Grasses capture carbon dioxide in a far more efficient manner than do other plants. They emerged around 60 million years ago, as lowering levels of atmospheric carbon dioxide forced plants to develop more sophisticated ways of capturing this increasingly rare gas. Without a reduction in carbon dioxide, grasses may never have developed, and humanity would consist of hunter-gatherers instead of sophisticated, fast-food consumers.

It's all about chance: stochasticism – if that's a word. Given that the probability of the same events occurring in the same order on another world must be vanishingly small. The probability that life will take the same meanders, oxbows and waterfalls is, therefore, highly unlikely. Consequently, the probability that an alien will walk off a spacecraft and say "Hi!" in broad American English seems slight. It may even be unlikely that the aliens we dream of are likely to exist. However, any alien will probably live on the land surface of a world bathed in its sunlight, held there by gravity, in an atmosphere that conducts sound. Thus, there will be obvious common features to us. What might these be?

Sense and Sensibility

Most organisms will have some way of sensing their environment. Without this they would probably be dead in no time at all. After all without a sensory system organisms can't find food, evade predators, or if plant-like, respond to changes in temperature, season and light. So, animal-like organisms will likely have some sort of visual system, auditory system and be tactile. They will need a brain and central nervous system of some description to coordinate activities if they are an animal larger than a few hundred cells in size. They will need some way of obtaining energy, either through capturing light or heat energy from their environment, much like a plant or bacterium on Earth, or they will consume food obtained from elsewhere, like an animal or fungi. Aside from this very little can be said about any possible alien life form. There are restrictions on what they can do, restrictions imposed by their environment, but in order to survive certain key features are obligatory. That's about it.

Therefore, the emergence of life on any suitable planet is very probable, but perhaps – and seemingly in contradiction to this confident assertion – we know very little about how this occurs. Finally, what such life would look like is difficult to say, but it will probably share common characteristics with organisms on our world. This is simply a virtue of the fact that life will have to adapt to conditions associated with the presence of liquid water, regardless of the planet it is on.

Oxygen: A Contradiction

Fossil evidence confirms an abundance of microscopic life by 1 billion years after Earth formed. Life was firmly established, with complex cellular chemistry extending to the process of photosynthesis. The fossilized remains of these organisms have been recognized in rocks that are 3.5 billion years old. Photosynthesis at this time was at a more primitive stage than that dominating Earth today. In today's plants, algae and cyanobacteria, water is split using light energy, liberating oxygen. At 3.5 billion years ago, the enzymes responsible for this chemical transaction didn't exist. The necessary twists and turns of mutation and selection hadn't furnished living organisms with the necessary equipment. Instead the volcanic gas hydrogen sulfide was split to liberate sulfur and hydrogen. Many bacteria still carry out the same reaction today. However, these are confined to relatively extreme environments, where oxygen is limited.

At 3.2 billion years ago, a type of bacteria – the cyanobacteria – emerged, the result of evolutionary pressures. The most abundant solvent on Earth is water. Hydrogen sulfide, by contrast, is restricted to volcanic environments, and those in which decomposition is occurring. Perhaps because volcanism was less extensive and hence hydrogen sulfide less prevalent, there came suitable selection for the use of a different fuel. Mutations within the gene that allowed hydrogen sulfide to be split caused its protein to select the more abundant molecule water instead of hydrogen sulfide. Admittedly, there is quite a lot more to it than that, but the principle holds true. A simple swap of sulfur for oxygen is all that separates these two compounds. To us they couldn't be more different. Hydrogen sulfide is a noxious gas that causes anesthesia in low quantities and death at higher ones. Water is only lethal to us through drowning, or through drinking it to excess. However, in terms of their chemistry they are rather similar.

How can we date the emergence of cyanobacteria and the use of water in photosynthesis? It's easy. At around 3.2 billion years ago Earth began to rust. The noxious byproduct, oxygen, began filling first the oceans then the atmosphere. In the oceans this highly reactive, and utterly poisonous, gas converted iron sulfide into sulfates and iron oxide. The sulfur-dependent bacteria were edged out as oxygen flooded their environment and drove them deeper in search of dark, oxygen-free water. Iron oxide – rust – precipitated onto the ocean floor in an orange rain, while strange domes of bacteria called stromatolites grew abundantly along the tidal shores of the ocean.

Why refer to oxygen as "noxious" or "poisonous" given its necessity for sustaining us and the gas we associate most with life? Oxygen is an odd gas. Oxygen is the second most reactive non-metal, after the voracious, but fortunately rare, fluorine. It poisons most life on Earth and stresses the rest into an early grave. This is an odd assertion, given our absolute necessity for it. Although oxygen is essential for all (or at least for the vast majority of) higher life-forms – those with a multitude of cells – the bulk of life on Earth is the invisible bacterial underworld. The majority of those either can't abide the stuff or have a take-it-or-leave-it attitude to oxygen. A cursory examination of our gut finds that greater than 99 % of the species that live there are more than happy without oxygen, perhaps surprising that this is inside an aerobic, or oxygen-dependent, organism – us.

In the environment, dig a few centimeters down into wet mud or rock, and oxygen becomes limited to less than 1 % of the abundance in the atmosphere. The organisms that are present live quite successfully without the poisonous gas. Most organisms live without the need for oxygen. Take it away and Earth would return to a primordial state with anaerobic (oxygen-loathing) microorganisms. It would be far from sterile.

Life does not need oxygen.

However, there is a caveat to this rather bold statement. Although the majority of species live merrily without oxygen, this gas, more than any other, seems essential for complex life.

As already stated, oxygen is very reactive. Metals are rapidly oxidized by it – they rust, in common garden language. The components of cells, like proteins or nucleic acids, are just as readily attacked by oxygen and its chemical minions as metals are. This often causes irreversible damage that can ultimately kill the cell. However, oxygen is as wonderful as it is detrimental. Its reactivity allows it to combine effortlessly with carbon compounds with a concomitant release of large amounts of useful energy (ATP).

In many, but not all, cells the dominant fuel is glucose. Through reactions, driven by a group of cellular proteins called enzymes, glucose is synthesized from carbon dioxide and water in plants and microbes, then regularly "burned" to liberate the same carbon dioxide and water. Of course, there are no flames. Instead, in tiny increments, glucose is first split in half and then progressively stripped of its hydrogen. What is left is a husk of carbon dioxide that we later expire. Hydrogen, liberated from its carbon jail, then participates in a series of convolutions that ultimately lead it to oxygen and the production of the energy-carrying molecule ATP.

Aerobic respiration is very efficient as a process, deriving a large fraction of the available energy as ATP. Combustion, although releasing more energy as heat, squanders this to the environment.

The process of oxygen-dependent, or aerobic, respiration is ubiquitous in all higher, or multi-cellular, organisms. Oxygen-hating anaerobic organisms derive energy by more unusual and effectively less efficient means. They succeed in life through other adaptations, including the use of considerably more fuel to achieve the same end. If each gram of fuel liberates less energy, you need to use more to survive and thrive. Oxygen affords multi-cellular life enough energy to carry out more complex tasks than mere survival.

Thus on Earth, there is a progressive evolution of life coupled to the rise of oxygen gas. For more than 1 billion years, the oceans absorbed the oxygen pouring out of cyanobacteria. By 1 billion years ago, the oceans were so stuffed full of the gas that it began to enter the atmosphere in significant amounts. Until this point Earth's atmosphere was predominantly nitrogen and carbon dioxide, with an uncertain, but probably significant, amount of methane, produced by anaerobic microbes.

As oxygen began to flood the atmosphere it first decimated any methane or ammonia present, vigorously oxidizing these gases to carbon dioxide and nitrogen, respectively. The rise of oxygen would then have two important effects on Earth. First, removal of abundant methane would make Earth's atmosphere less able to hold onto heat. Methane is a highly effective greenhouse gas, with ten times the heat-harvesting power as carbon dioxide. Loss of methane would continue a process of cooling the planet down, something driven by the steady consumption of carbon dioxide in the oceans.

The second effect is related to the fact that oxygen also combines with itself under the influence of ultraviolet light. The product is the poisonous gas ozone. Again, there is an apparent contradiction here. Ozone is a highly reactive form of oxygen. Breathe it in and die. But, place it high enough in the atmosphere and it forms a protective blanket, screening out the Sun's harmful ultraviolet light. Ozone breathes life into Earth by absorbing ultraviolet radiation, which can cause serious harm to any organism that is exposed to it.

Thus, at around 1 billion years ago, the stage was set for some explosive transformations on Earth. Oxygen gave life two important breaks. It protected the surface from harmful ultraviolet light. Life could then spread from the insulating oceans onto the exposed surface of the land. Undoubtedly, the soils and rocks of Planet Earth were thriving with microbial life. But the surface was probably as sterile as the surface of Mars is today. Oxygen removed an impasse caused by ultraviolet light, allowing aerobic life to spread microscopically across the land's surface.

If one could observe this from space, the surface might slowly change from murky shades of brown to green as photosynthesizing bacteria spread. Meanwhile, the rise of oxygen was allowing organisms to produce more energy and carry out progressively more complex tasks. Cells could begin to clump together, affording the opportunity for delegation of biological tasks. Although not strictly multi-cellular organisms, these aggregations of cells were able to carry out intricate operations, beginning a protracted journey down the road to complex, higher life.

Land also has an important role in evolution. In the oceans conditions are relatively constant. There is variation in temperature, light and nutrients, but the temperature range is relatively muted, and the amount of light varies in synchrony with depth. Given these fairly slow and steady changes, organisms adapted to their aquatic environment don't need to worry too much about change. By contrast, conditions on land vary, often significantly over short periods of time. Temperatures vary, sometimes in extreme ways, between day and night; water can be limiting; light varies considerably over the course of a day; and over longer periods of time plate tectonics – the slow movement of a planet's crust – means that habitats change over millions of years as they drift from one latitude and longitude to another.

Early on in Earth's history land was undoubtedly a limiting factor. Most of the planet was hidden under hundreds or thousands of meters of water. As plates shuffled around, driven by mantle convection, land was driven up above the water's surface by volcanism, through the collisions of portions of crust and also by the steady growth in the amount of buoyant granite rock. The literal rise of land was utterly essential to the later evolutionary steps needed for the development of complex mammalian and avian life. Without mantle convection, the planet would probably be limited to aquatic organisms – possibly pretty but hardly likely to form the basis for intelligence.

Multi-cellularity

Somewhere between 2.5 and 1.5 billion years ago, on Earth a new type of cell emerged that could work with oxygen. This was the eukaryote. We are eukaryotes, as are plants, fungi and a host of weird and wonderful organisms. What sets eukaryotes apart from their bacterial (or prokaryotic) siblings is the complexity and functionality of their cells.

Without being disrespectful to the microbiologists, prokaryote cells are relatively simple structures. Their genetic material sits in the same compartment as the enzymes that keep the cells alive. The cell is bounded by a single fatty membrane, in turn enclosed by a wall of varying composition. All bacterial cells share this structure, although a few have done away with their cell walls.

Eukaryotic cells, by contrast, store their chromosomes – the home of genes – in a nucleus. This is itself a complex fat and protein structure, drilled with holes that allow the genes to communicate with the rest of the cell. Other structures inside the cell manufacture and deliver proteins, carbohydrates and fats from raw materials; while others are designed to degrade attacking pathogens or worn-out portions of the cell. Finally, there are the mitochondria of plants, animals and fungi and the chloroplasts of plants and some other animal-like cells (Fig. 7.4).

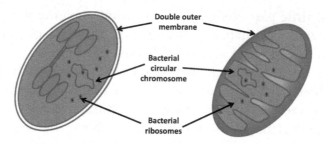

Double outer membrane

Bacterial circular chromosome

Bacterial ribosomes

FIG. 7.4 Crude schematic of the eukaryote cell's chloroplast (*left*) and mitochondrion (*right*). Various features within these cellular organelles confirm that they were once free-living bacteria. As well as bearing physical structures that match their free-swimming bacterial counterparts, various other features (indicated) are clearly bacterial and are quite unlike the corresponding features of the cells in which they are now found. The process by which they became incorporated into the cell is known as endosymbiosis

Mitochondria are our powerhouses. They are the cellular structures that use oxygen and allow us to grow, manipulate tools, think and write books. By using oxygen, mitochondria generate the vast bulk of the energy we get from glucose. Without mitochondria we'd still be dwelling in mud, contemplating nothing.

In plant cells, chloroplasts busily take energy from the Sun and use it to split water, liberating oxygen. Indirectly, they combine the hydrogen with carbon dioxide, manufacturing the glucose and other raw materials animals take for granted (Fig. 7.5).

However, mitochondria and chloroplasts are not what they seem. Once they were free-living souls, microbes that lived in the oceans or in mud along the edges of waterways. The chloroplasts were cyanobacteria-like organisms, cells we still find today, almost unscathed by more than 3 billion years of evolution. However, probably not through choice but by consumption, some of these free-living cells became meals for early eukaryotes. Fortunately for these bacterial dinners, the eukaryote is often a clumsy eater.

Even today bacteria often evade digestion inside their membranes, escaping instead to take up residence. A little over 1.5 billion years ago a type of purple bacterium that had evolved to use oxygen moved in and became the mitochondrion. A few 100 million years later in another chance event, the cyanobacterium, or its ancestor, moved in and took up tenancy. Collectively, biologists refer to these internal cellular structures as "organelles," and the

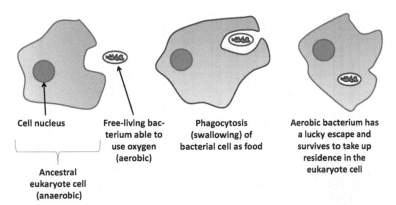

Cell nucleus

Free-living bacterium able to use oxygen (aerobic)

Ancestral eukaryote cell (anaerobic)

Phagocytosis (swallowing) of bacterial cell as food

Aerobic bacterium has a lucky escape and survives to take up residence in the eukaryote cell

FIG. 7.5 An outline of endosymbiosis. A free-living, oxygen-utilizing bacterium is engulfed (phagocytosed) by a larger, anaerobic eukaryote cell. In most cases the bacterium is digested as food. However, in many cases the bacterium survives. As it provides a clear benefit, in the form of extra energy generation, to the host cell, the cell retains and establishes a symbiotic relationship with it. Over time genes move from the bacterium to the nucleus, and the two organisms effectively merge. The process of gene flow from mitochondria and chloroplast to nucleus is still ongoing and has been measured in plants

process by which they became permanent additions to the cell as "endosymbiosis." In etymological terms endosymbiosis means internal partner, or symbiotic organism (Fig. 7.6).

How do we know that the process of endosymbiosis explains the origin of the chloroplast and mitochondrion? Not only have these symbiotic organisms left a trail of cellular structures inside their cells, they have also left genes, molecular fingerprints that point to their activities. The mitochondrial genome – its DNA – in human cells only codes for 13 proteins and a handful of other functional RNA molecules. The precise DNA sequences of these genes (the order of their A, C, G and T units) ties them to their free-living bacterial relatives, as do specific features of the cells, such as ribosomes, which are clearly bacterial and not human.

Similarly, the more recently acquired chloroplast has a relatively large, circular genome containing hundreds of genes. Again, the genetic organization of these is virtually identical to those in cyanobacteria, in some cases right down to the fine print. Like their mitochondrial cousins, the structure of the ancestral cyanobacterial cell is often well preserved, in a few cases even retaining the original bacterial cell wall.

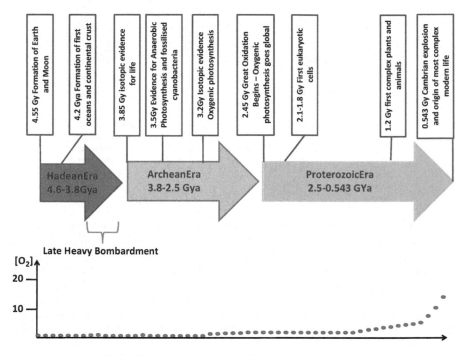

FIG. 7.6 Some of the key moments in the rise of life on Earth matched to the changes in the level of oxygen in the atmosphere

If the chloroplast just happened to resemble a cyanobacterium, why would it need a bacterial cell wall in a plant cell? Why would the order and structure of its DNA match the free-living bacterium? There is no valid alternative explanation. By chance alone they would be different, and they are not.

We also know that this origin is true because we can observe the same process occurring today. There are some marvelous examples of algal and animal cells that take up cyanobacteria from their surroundings. Instead of simply digesting these meals, they retain them, stealing the glucose and other compounds they are producing using the Sun's energy. Interestingly, this includes the malaria parasite, *Plasmodium falciparum*.

You might ask why there would be a structure involved in photosynthesis in the cell that does not really carry out the process. There is no real reason why a parasite of mammals would have such a feature. The answer might be because it retains much of its original cyanobacterial structure. Photosynthesis isn't part of its life cycle. However, it is lucky for us that it does.

This fossil structure carries out some unlikely and un-mammal-like functions. These functions could well turn out to be this lethal parasite's Achille's heel. They could well make it vulnerable to drugs that won't harm us.

The Links Between the Biosphere and Geosphere

The rise of complex life on Earth was an intricate, long-winded process lasting nearly 4 billion years. Undoubtedly, key steps towards this include the formation of an oxygen-rich atmosphere, which could support the activities of large organisms and protect them from ultraviolet light. Plate tectonics at the very least is a beneficial feature of any world on which life emerges, and more than likely is a downright prerequisite. Plate tectonics modifies the levels of carbon dioxide, regulating temperature within the bounds set by the liquidity of water and the conditions necessary for photosynthesis. Plate tectonics generated dry land by raising portions of crust above the surface of the global ocean, where the creation of new environments could spur on the processes of evolution. Plate tectonics delivers cold crustal material into the mantle, where it stirs convection within the underlying liquid, outer core (Chap. 4). An absence of global plate tectonics could be the reason the atmosphere of Venus is now battered by the solar wind, while ours nestles sweetly within a protective magnetic field.

Is a moon important for the development of life? Certainly, our Moon helps stabilize the tilt of the planet, giving fairly constant seasonal differences over millions of years. However, we have good evidence that even the Moon's gravitational pull was insufficient to prevent large and relatively rapid movement of the continents in the past. On Earth, such true polar wander appears to have been caused by the slow motion of continental crust, bringing large masses of continent to polar locations.

In terms of Earth's spin it is more stable to have the continents arranged near the equator, where they are pulled upon by the spin of the planet. Place the continents near the poles, and the spin of the planet becomes unstable because of the distribution of mass. The uneven distribution of mass pulls the surface in a direction against the centrifugal force caused by our planet's spinning motion. The crust then slides over the mantle, until the greatest concentration of mass resides once more near the equator, and stability is restored. The process takes millions of years and is far from catastrophic. There is nothing to suggest these relatively rapid movements did life any harm in the past and may have spurred on further evolutionary innovation.

Planets orbiting within the habitable zones of red dwarfs are tidally locked to their parent star, so there is no requirement for an accompanying Moon to stabilize the planets rotation. The gravitational pull of the star does the work through the tides it raises. However, the vagaries of plate tectonics may mean that periodically the surface does suddenly move through true polar wander. Remember "suddenly" is a relative term. This would expose new territories to different environmental conditions. At a slower pace, plate tectonics may be the prime mover of environmental change, true polar wander or not. This is not to say that true polar wander may not be without cost to life on a tidally locked planet. Take a look at our Moon and its distribution of crust. We are familiar with the "man on the Moon" – the dark lunar maria. These are, almost exclusively, found on the Earth-facing side, with the far side predominantly light, lunar highlands. This may not be coincidence.

Modeling work by Oded Aharonson and colleagues at the California Institute of Technology suggests that the pull of Earth has acted to orientate the Moon in this manner. Their work suggests that tidal forces will tend to concentrate highlands on one side and the denser basins on the other. The chance of the basins facing Earth was 50:50 if the Moon slowed its rotation quickly. However, a slower rate of decelerated more slowly there was a higher probability that the maria would face Earth. If this holds true for tidally locked planets perhaps half or more will tend to aggregate their greatest concentration of continents on the dark, sunless side, while the other 50 % orientate it on the bright sunlit side. Periodically plate tectonics might shuffle the deck. However, whenever this process ceases there will be an element of pot luck as to whether the continents will have carried their retinue of living organisms over into permanent darkness, thus eliminating any complex life living there. This is yet another factor that must be taken into account when considering the habitability of crimson worlds.

So, for life the best we can say is that we need a sizable world with a gravitational pull strong enough to sustain a thick atmosphere that allows the presence of liquid water. In addition we probably require enough mass to retain the heat necessary for sustained plate tectonics. It's not clear if anything else matters. Moons may be a pretty aside, but apart from moments of romance may have little practical benefit for life on any planet orbiting within the habitable zone of a red dwarf. Indeed, it isn't even clear if a large satellite could stably orbit a planet that was tidally locked to its parent star. More likely than not, tidal forces from the parent star would strip it from its orbit very soon after its formation.

Identifying Living Worlds

How can we identify potentially habitable worlds around red dwarf stars? The dual methods of spectroscopy and polarimetry offer solutions. Gather enough light from your distant world and certain clues may emerge in what you see.

The simplest method for obtaining such information is the transit. When a planet crosses the face of its star there are two brief intervals at the beginning and end of the transit when the light of the star passes through the atmosphere of the planet. Gases within the atmosphere absorb specific wavelengths of radiation, and these emerge in the stellar spectrum of the star. With sufficient resolution these minute absorptions can be distinguished from the stellar background.

A few such spectra have been obtained, initially by the Hubble Space Telescope, and latterly by the Kepler instrument. Primarily spectra are of hot Jupiters – for the simple reason that these planets orbit closest to their stars and thus transit their star frequently. Moreover, the shadow of the planet occupies a proportionately larger area of the stellar photosphere in the transit because the Jupiter-mass planet is larger. A small subset of hot Jupiters are so intensely irradiated that their atmospheres are boiling away into space. This process generates a cometary wake of material that streams away from the planet and star into interstellar space. These engorged atmospheres were an easy target for Hubble, and the first atmosphere to be sampled was that of HD 209458b.

More challenging was the carbon monoxide-rich atmosphere of TrES-1 – another hot Jupiter, but one cool enough and massive enough to hold onto its gases.

More challenging still will be the potentially habitable Earth and super-terran mass worlds orbiting in the stellar habitable zones. For one thing they are much smaller worlds. However, this is only part of the problem. Their atmospheres are less extensive, hug the planets more tightly and most importantly are attached to planets that make less frequent transits of their stars. Orbiting further afield means their orbits may be measured in weeks to months, depending on the mass of the parent star and the breadth of its habitable zone. Moreover, with less intense irradiation to illuminate the planetary atmosphere, absorption lines are less distinct.

There is one more problem with the use of the transit method when we look at habitable worlds orbiting K- and M-class stars. These low mass stars are themselves very cool. This means that their atmospheres are clogged with a rich pencil-work of absorption features. M-class stars, in particular, have very complex spectra (Chap. 3), and disentangling the absorption features of the star from those of its planet will be challenging.

However, two new methods have emerged that may be usable. The first was described in Chap. 1 – the use of the CRIRES spectrograph by Matteo Brogi's team at the VLT in Chile. This was used to determine the spin of τ Boötes b. To quickly recap, τ Boötes b did not transit its star, but using the very sensitive infrared spectroscope fitted to the VLT, astronomers isolated the emission of radiation from hot carbon monoxide gas in the atmosphere of the planet. Carbon monoxide is absent in the atmosphere of the host star but is abundant in the atmosphere of the planet. By observing the system with high resolution over the course of several days the orbit of the planet and its spin could be determined by monitoring the motion of a hot spot within the atmosphere of the star. Brogi's team were able to pinpoint the location of the noxious monoxide gas within the lower atmosphere of the planet. With improved sensitivity the same method could be applied to other planetary systems to sample the atmosphere of their worlds.

In 2012 a novel technique was published: the use of polarimetry to sample the surface and atmospheric chemistry of Earth using Earthshine. Sunlight passing through Earth's atmosphere and reflected from Earth's surface is called Earthshine. This secondary radiation has been thoroughly modified by its passage through our atmosphere. Such light contains straightforward spectroscopic information, but it is also altered. It is polarized.

Polarization comes in two flavors – linear and circular polarization. In essence the light is bent in directions specific to the chemicals with which it interacts. The extent of polarization thus serves as a chemical fingerprint that can be read. Polarization is more sensitive than spectroscopy, and the unique bend or twist imparted on light by each chemical allows the effect of planetary atmospheres to be separated from the influence of the parent star.

As a proof of principle, Michael Sterzik used the European Southern Observatory to analyze Earthshine reflected from the Moon. These spectro-polarimetric observations identified a number of key planetary features such as the proportion of the surface covered by oceans and vegetation, as well as the extent of cloud cover. In principle this type of analysis could be applied to radiation reflected from the surface or atmosphere of any distant planet. The expectation is that this technique will be applied in the not too distant future, allowing us to sample both the composition of the planetary atmosphere and its distribution of clouds. With this handle we might eventually be able to determine the weather on Gliese 581d or g and decide whether we might need a umbrella when we visit.

Astrobiologists also hope to use spectroscopy to measure the presence of vegetation through a subtle feature bearing the seemingly politically leaning phrase "the red edge" (Chap. 8). Terrestrial plants have a specific pattern of absorption of visible radiation. Beyond 700 nm (the red end of the visible spectrum), absorption into the near-infrared end of the spectrum is rapidly muted. This gives rise to a "red edge" in the absorption spectrum for the planet's surface as a whole. Should other planets have plants like ours – a fairly big "if" – then we might see the same or similar absorption "edges" as seen on Earth. However, it assumes no further absorption of radiation beyond "the edge," or that organisms won't have evolved to carry out the process using longer wavelength radiation.

This proposition is somewhat dubious, since we know from Earthly experience that evolutionary processes tend to fill available niches, given the opportunity. The planet of a red dwarf will be bathed abundantly in such long-wavelength infrared radiation. Therefore, one might assume that any oxygenic photosynthesis on the planet of a crimson star may lack a distinct red edge, or one that is shifted more into the infrared portion of the electromagnetic spectrum. At present we are unclear as to whether oxygenic photosynthesis is possible at longer wavelengths (Chap. 8), but we shouldn't rule it out just yet.

With future propositions, such as ESA's Darwin or NASA's Terrestrial Planet Finder, we should be able to thoroughly scrutinize the light from transits or from reflection by these methods. An exciting decade awaits – funding permitted.

Conclusions

The rapidity of life's emergence on Earth suggests, in turn, that this is a highly probable outcome. Early Earth was a brutal place, subject to intense ultraviolet irradiation, frequent bombardment and an atmosphere we would find utterly noxious. Yet, within 50 million years of the close of the period of heaviest bombardment, the chemical signature of life emerges and is preserved in our rocks. Within 300 million years of this juncture, life that we might recognize appears and is fossilized.

It is reasonable, therefore, to assume that life readily arises and should be able to do on any other world given the opportunity. This term "opportunity" sounds vague, and perhaps it is. We know that life has adapted to occupy a myriad of environments on Earth and that seemingly harsh environments are no obstacle to its rise. We then assume that it will do so elsewhere. Given that life can be reduced to two basic principles, information storage and evolution, this assumption is more than reasonable.

We assume that some biochemistry is likely to be universally common, such as the use of the universally abundant amino acids. However, the mechanisms of information storage could be locally unique, unless we assume panspermia.

The mechanism behind evolution is uniquely simple – the failure to copy information accurately. An error rate of 1 in a 100,000 is high for biological organisms, but mistakes are a part of life. Once you have a copying system that is error-prone, biologists would probably say that life is inevitable – and complex, cellular life at that.

We can infer no more from common experience. Everything else is a gedanken experiment, but what an experiment. In varied but perhaps not limitless environments, life could elaborate into virtually endless forms. With water, energy, organic compounds and the time and scope for experimentation, the universe is your oyster – or lobster or tentacled slime-filled, sentient, photosynthesizing avian marsupial. Our assumptions are limited not by everyday experience but by our imaginations. Dream on.

Part II
Life Under a Crimson Sun

8. Red Dwarfs, PAR and the Prospects for Photosynthesis

Introduction

Red dwarfs pose some unique challenges to any life that attempts to take a foothold on one of their worlds. The bulk of the radiation received at any orbiting planet's surface is low energy infrared, rather than the dominantly visible wavelengths received at the surface of Earth. Water needs a critical input of energy if its bonds are to be broken. Without this energy hydrogen cannot be liberated from the molecule and made available to convert carbon dioxide into carbohydrate. As the byproduct of this reaction is oxygen, the atmosphere of any habitable planet orbiting it will be bereft of this molecule.

However, water is not the only substrate for these reactions. Hydrogen sulfide can also be used, and this requires less, though still significant, energy. In this chapter we explore the needs of photosynthetic organisms on Earth, the limitations imposed upon the process on any red dwarf planet, and the potential for maneuvering the biochemical reactions in organisms to allow oxygen to be produced in abundance.

A Primer on Photosynthesis

Plants on Earth are rather obviously a green color – at least for most of the year. The green coloration is predominantly due to two pigments, chlorophyll a and b. Chlorophyll b absorbs more energy in the far blue and red ends of the spectrum than chlorophyll a and is somewhat greener in color than chlorophyll b. By absorbing blue and red light, chlorophyll molecules contribute to the healthy, green hue of living plants. So pervasive is the green of

D.S. Stevenson, *Under a Crimson Sun: Prospects for Life in a Red Dwarf System*, Astronomers' Universe, DOI 10.1007/978-1-4614-8133-1_8,
© Springer Science+Business Media New York 2013

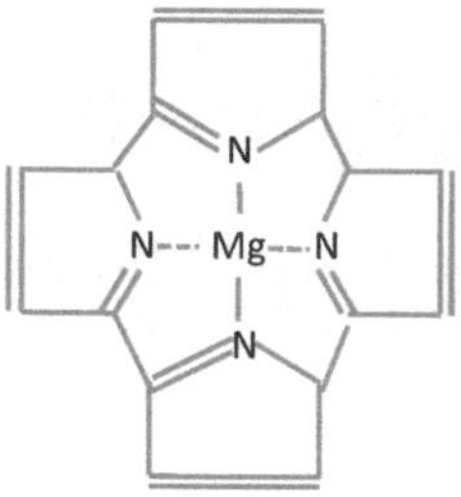

FIG. 8.1 A simplified view of the central core of the chlorophyll molecule. Light striking the magnesium ion in the center removes electrons (oxidizes it). These electrons are used to carry energy from the Sun to the energy-using machinery of the cell. In plants and some bacteria, the electrons are replaced with those from water, which allows it to be split into oxygen and hydrogen

chlorophyll that we unconsciously associate the rebirth of spring with the greening of the land. On Earth, green means life. In times of recession, politicians pontificate on the presence or absence of the "green shoots of recovery," which may or may not be evident depending on your political bias.

Chlorophyll is a wonderfully complex organic molecule, consisting of four rings of carbon, nitrogen and hydrogen, encasing a single magnesium ion (Fig. 8.1). Chlorophyll a and b differ slightly in the structure of a tail that is attached to this ring structure, and this difference explains the difference in the absorption of light by each molecule.

When light of an appropriate wavelength strikes chlorophyll, some of the electrons surrounding the magnesium ion can receive enough energy to jump off. These "excited" electrons can be used to instigate more exciting chemistry within the cell.

Chlorophyll doesn't simply slop around the cytoplasm of the cell – the liquid goop that fills the bulk of the cell. Instead it is confined by proteins to the fatty membranes of the chloroplast. Here, it is organized in a unique way that allows those excited electrons to flow effortlessly from protein to protein and ultimately onto a compound that can convert carbon dioxide into glucose. Along the way, the electrons lose energy. The really ingenious aspect of this process is that this energy is used by the chloroplast to synthesize the energy molecule ATP (Chap. 7). This ability necessitates the presence of a series of fatty membranes within the chloroplast. These membranes form an impermeable barrier to the movement of water and ions. This arrangement allows the chloroplast to control how energy is managed, ultimately channeling the capture of light energy into a more useful biological commodity, ATP. The chloroplast thus serves as a source of chemical energy for the plant and by those organisms that feed upon them.

Photosynthesis can be reduced to a few key steps. In essence the first stage is when chlorophyll is struck by light of a suitable wavelength and it becomes excited, losing electrons. During the subsequent, fast stages, these electrons pass through a series of molecules and onto carbon dioxide, converting it to glucose. The leftover highly oxidized[1] chlorophyll molecule then rips electrons from water, smashing it into its components, hydrogen and oxygen.

In the next stage, oxygen escapes into the atmosphere, while the hydrogen remains, dismembered into free ions (protons) and electrons. The electrons that are released from water replace those lost from chlorophyll through the action of light. Meanwhile the hydrogen ions build up on one side of an internal fatty membrane within the chloroplast, much like water behind a dam.

The key to the capture of useful energy is this dam. Hydrogen ions are trapped on one side of the membrane, but most importantly they can only cross the membrane and even out the concentration differences through specialized channels – an intricate, beautiful enzyme called ATP synthase.

How does ATP synthase convert energy stored in the gradient to the useful chemical energy in ATP? The system is utterly ingenious. The enzyme that produces ATP works just like the generator in a power plant. In a generator a large magnet spins inside a coil of wire, inducing an electrical current in the wire. In an amazing microscopic reflection of our generators the hydrogen ions flowing relentlessly through a portion of the ATP synthase enzyme cause the core of the protein to spin at high speed (Fig. 8.2). As this spins, parts of the protein changes shape pushing the building blocks of ATP together. This is beautiful system is utterly ingenious. It is simply amazing that we generate electricity in a manner equivalent to the way our cells produce ATP.

---

[1] In chemistry oxidation means the loss of electrons and reduction is the gain of electrons. The mnemonic "OILRIG" is used in teaching to remind students of this: **O**xidation *is* **l**oss of electrons and **r**eduction *is* **g**ain of electrons.

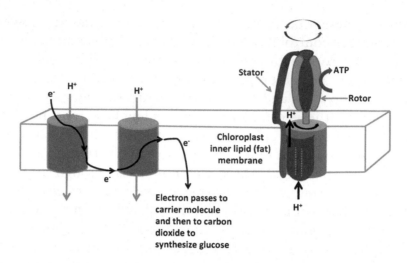

FIG. 8.2 The marvelous ATP-synthesizing dynamo of chloroplasts and cyanobacteria. As electrons flow from chlorophyll to carbon dioxide hydrogen ions (H⁺) build up on the outside of the inner chloroplast membrane and are only allowed back through this impermeable fatty layer via the ATP-synthesizing enzyme. As they pour through a protein channel they cause the rotor to turn. Instead of electricity, this turning action produces ATP, the energy currency of the cell

Within the chloroplast membranes chlorophyll molecules are associated with an army of additional compounds that allow plants to harvest energy at wavelengths other than the reds and blues preferred by them. For example the oranges, yellows and reds of carrots and peppers are down to the presence of molecules called carotenes. These capture blue light along with some reds and greens. Additional molecules capture light at other visible wavelengths, and together they trap much of the Sun's visible energy (Fig. 8.3).

Light capture drives the movement of electrons into a chlorophyll molecule at the heart of the large protein complex. This is known as the light-harvesting complex. This concoction of proteins and pigment molecules is unique to photosynthesizing organisms, and ensures that plants, and their bacterial ancestors, are able to capture the broadest possible range of wavelengths of light. This, in turn, maximizes their ability to capture carbon dioxide from the air. The light-harvesting complex bequeaths a plant both the freshness of green in the spring and the magnificent, changing hues of autumn.

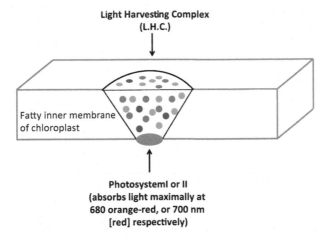

FIG. 8.3 The structure of the light-harvesting complex in the chloroplasts of higher plants or cyanobacteria. A collection of light-absorbing pigment molecules nestles with protein in a structure in the inner membrane. At the base of this lies the photosystem – a specialized protein containing hundreds of chlorophyll molecules and accessory pigments that absorb light. Electrons liberated from the L.H.C. funnel through the different pigment molecules ultimately ending up in the reaction center

The colors of plants are specific to both their underlying genetics and certain environmental cues. On tidally locked worlds with one season, we may expect a permanent hue to be dominant. However, given even the slowest rotation and a significant angle of tilt, these planets, too, may reverberate with the splendor of seasonal change.

Action Spectra and Beyond

The amount of photosynthesis a plant can carry out varies significantly with wavelength. Simple experimentation can be used to compare the rate of photosynthesis with the irradiating wavelengths of light a plant or a bacterium receives. Monochromatic sources (usually light emitting diodes – LEDs) may be used to deliver light in particular ranges of wavelength. The rate of photosynthesis can readily be measured by quantifying the amount of oxygen the organism releases in a given time when illuminated.

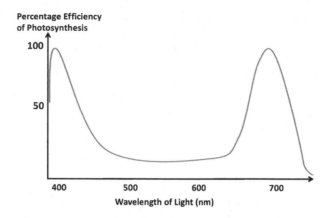

FIG. 8.4 The variation in the rate of photosynthesis with the wavelength of green plants on Earth. This "action spectrum" relates wavelength to the rate. The maximum absorption of useful energy occurs at the red and blue ends of the spectrum. Plants are green because little light is absorbed at green wavelengths

By comparing wavelength with oxygen release, an action spectrum is produced (Fig. 8.4). In essence this is a simple means of quantifying the effect of light on a photosynthetic organism.

Most higher plants show a pattern similar to that shown in Fig. 8.4. The rate of photosynthesis is highest at the blue and red ends of the photosynthetic spectrum and lowest in the green portion. Given that plants and photosynthesizing cyanobacteria are green, this is to be expected. Most green light is reflected. Without absorption, the plant cannot capture the Sun's energy. There is a rapid fall off in absorption in the infrared portion of the spectrum, giving rise to the so-called "red edge." Sounding like some form of troublesome political persuasion, the red edge may be used to distinguish life-bearing habitable worlds in future. By analyzing variations in light reflected from planetary surfaces, across the red and infrared portions of the electromagnetic spectrum, astronomers may be able to observe the signatures of Earth-like biospheres.

However, this is not the whole story. Light is captured by a complex of different pigments and proteins, not individual pigment molecules. The core of each light capture complex absorbs maximally at two particular wavelengths in plants: 680 and 700 nm. The remaining wavelengths are absorbed by the accessory pigments, which then deliver the energy to the core of each complex. Consequently, each complex has a range of wavelengths over which it can absorb useful energy. Indeed, this range extends into the near (the shortest wavelength portion of the) infrared.

In some bacteria, the maximum absorption occurs in the infrared portion of the spectrum. Plants, too, will absorb radiation in the near infrared. Yet, plants are a little more choosy, needing a little more if they are to capture energy efficiently. Where radiation is absorbed in the near infrared, little photosynthesis occurs unless bluer light is also provided to compliment the infrared flux. Therefore, the observed rate really depends on the appropriate combination of wavelengths, rather than specific wavebands. The question is, does a cool red dwarf star supply the types of radiation needed by our land plants? Could we transplant Earth to a suitable orbit around a red dwarf and see life forms such as us thrive?

Table 8.1 shows how the wavelength peak of light varies with stellar mass and surface temperature. The interesting point is that the coolest, lowest mass red dwarfs primarily release infrared light, which is left of the peak of maximum absorbance for terrestrial land plants and cyanobacteria. However, even for the coolest stars, as much as 10 % of the visible radiation has a wavelength shorter than that required for land plants to flourish. This is well within the tolerance range for terrestrial plants, which typically photosynthesize effectively even when light intensities are low, either because of shading or because of cloud cover. Therefore, even under the light of a crimson sun, terrestrial plants – those from Earth – would probably survive and prosper (Fig. 8.5).

Wavelength and the Products of Photosynthesis

On Earth many species of bacteria absorb light at wavelengths extending well into the infrared. For example the exquisitely named purple non-sulfur bacterium, *Rhodospirillum rubrum* and the equally challenging tongue-twister *Rhodopseudomonas capsulata* have pigments that absorb infrared radiation at 870 nm. Other bacteria contain the molecule bacteriochlorophyll b, which absorbs light maximally at 960 nm – well into the infrared. However, these organisms do not split water to release the vital hydrogen

Table 8.1 A comparison of some stars, their radiation and the distances to the stellar habitable zone

Spectral type	G2 (The sun)	M0	M1 (Barnard's star)	M2	M3 (Gliese 581)	M4	M5	M6	M7	M8 (Proxima centauri)
Blackbody temperature (K)	5,860	3,850	3,720	3,580	3,470	3,370	3,240	3,050	2,840	2,640
Orbital radius (A. U.) for Earth-like insolation	1.00	0.26	0.23	0.19	0.15	0.12	0.10	0.07	0.04	0.02
Orbital radius (stellar radii) for Earth-like insolation	214	56	48	41	33	27	21	14	9	5

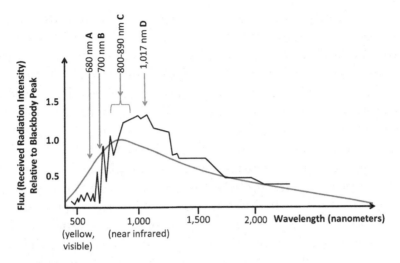

FIG. 8.5 The effective radiation released by a 0.1 solar mass red dwarf compared to its idealized, or blackbody, curve. Listed alongside this are four labels (A–D). These represent, respectively, the peak absorption of light by Photosystem I, Photosystem 2 of plant chloroplasts, and the bacterial photosynthetic pigments bacteriochlorophyll a and bacteriochlorophyll b. The latter two absorb maximally in the infrared portion of the spectrum, rather than the visible wavelengths of light. The so-called "Photic Zone Window", where light penetrates water, lies near 500 nm

gas needed to drive the production of glucose. Instead they use the lower energy of infrared radiation to split molecules, such as hydrogen sulfide (H_2S). Importantly, the radiation emitted by red dwarf stars is primarily in the infrared portion of the spectrum – exactly that needed for non-oxygenic photosynthesis.

On Earth infrared radiation is sufficient for photolysis (splitting) of hydrogen sulfide but not water. Therefore, although terrestrial plants can do the job at a lower rate with infrared radiation, conceivably oxygenic photosynthesis may not evolve at all on worlds of crimson stars. If this is true, there will be an inevitable and unfortunate consequence for life on any habitable world orbiting a red dwarf star. Splitting hydrogen sulfide produces sulfur and, indirectly, water; therefore, an Earthly biosphere based on this type of photosynthesis would lack oxygen. Since it seems likely that the ferociously reactive element oxygen is required for the development of complex organisms, life on a world lacking suitable photosynthesis might never make it past the use of the microbial template. Complex life may be precluded.

Oxygenic photosynthesis evolved a few hundred million years after sulfugenic photosynthesis and required considerably more complex machinery. The evolutionary delay is most likely a consequence of the greater requirement for energy to split water rather than hydrogen sulfide. Cyanobacteria developed the process by linking two pre-existing but distinct energy-capturing processes. Once these systems were linked the bacterium could grab enough energy to split water and liberate oxygen.

This step only occurred once in our planet's history. This might suggest that the probability of it occurring is low and hence unlikely to happen. If this were true where the available energy was lower, the low probability might preclude it from happening at all. Alternatively, it may have arisen only once on Earth because it was so successful that all the available niches for photosynthesizing organisms were quickly taken. This would have prevented duplicates from subsequently arising. It isn't at all clear which of these explanations is true. If it is the latter, we may not have much to worry about.

Life will take the easiest route when given the choice. Thus the lower energy non-oxygenic photosynthesis evolved before the more energetic oxygenic reactions.

Can we envisage a mechanism that will allow photosynthetic organisms on crimson worlds to carry out oxygenic photosynthesis despite the more limited energy supplies? To get around the energy shortage, life on a planet orbiting a red dwarf might use some form of energy storage to allow it to capture enough energy to split water. Think of powering the world's most powerful laser using the national grid. You can't simply plug in a trillion watt laser and click a button. Instead, banks of capacitors store increasing amounts of charge. Once enough has been stored the laser is fired.

Analogously the interior of the cell could be rigged to store sufficient charge to allow the splitting of water. Photons of light might be captured sequentially, driving greater and greater oxidation of some intermediary. At a critical point there is sufficient power to enable the splitting of water and the release of oxygen gas. Such processes would require additional combinations of light-harvesting components that coalesce in sufficient numbers or packets of lower energy. This would require a reconfiguration of the cells' ingredients so as to properly insulate the critical components of the light-harvesting complex – essential to ensure enough capacitance is stored to break water's tougher bonds.

All of this is theoretical. It is utterly unclear whether combining packets of very low energy is possible in such a way. Certainly, there are no precedents for such biochemistry on Earth. Yet we must remember terrestrial organisms don't do this simply because they don't need to. All we can say is that life is robust. Our senses tell us that evolutionary processes are adept at circumventing difficult circumstances. We can no more rule it out elsewhere than we can observe it on Earth.

However, let's assume that infrared light can drive oxygenic photosynthesis. There is one final, and perhaps unexpected, constraint on its operation. The atmosphere may not be transparent to infrared radiation at the correct wavelengths. On Earth, our atmosphere absorbs all infrared radiation short of 1.5 μm (millionths of a meter). Water absorbs the bulk of this, but there are additional peaks in absorbance by methane and carbon dioxide, as well as a host of other compounds. For instance, infrared radiation is efficiently absorbed by methane at wavelengths longer than 1.6 μm (millionths of a meter). Since terrestrial living organisms produce methane in abundance under anaerobic conditions, its presence on the worlds of crimson stars may be likely. With enough of this gas in the atmosphere critical wavelengths of infrared may be unavailable for photosynthesis. Clearly this would restrict the development of photosynthesis in the first place.

Other compounds, such as carbon dioxide and water vapor, may also restrict what wavelengths are available for active photosynthesis. It could well be that the rise of methanogenic (methane-producing) life actually prevents the development of oxygenic photosynthesis driven by infrared radiation. On Earth this problem was prevented by the availability of ample visible radiation, but perhaps on a world flooded with infrared radiation, such a escape route may be blocked.

Might the universe's most abundant and smallest stars host only anaerobic life, while the K- and G-dwarfs turn out to be the only abodes of oxygenic life? Only future observations will tell.

However, let's not be so pessimistic. If oxygenic photosynthesis can be cracked on such a world, the rise in oxygen would soon clear the fog of methane and drive further biosphere development. Any oxygen will readily react with atmospheric methane in a matter of years, oxidizing it to carbon dioxide and water. This would soon eliminate the restriction placed upon life and allow it to prosper.

Looking more widely, why is it that photolysis (splitting with light) of water dominates the source of usable hydrogen on Earth? Why don't we have a biosphere dominated by hydrogen sulfide, perhaps some temperate version of Jupiter's moon, Io?

This comes down to the stars themselves. The underlying physics of stars imposes important restrictions on the use of hydrogen sulfide or water by microorganisms. Put simply, stars synthesize far more oxygen than they do sulfur. This is a reflection of the increased complexity and demand in the manufacture of sulfur by nuclear fusion. Oxygen is made by all stars over approximately 0.5 solar masses. This represents greater than 40 % of the stars in the universe. Thus oxygen is an abundant gas in the universe. By contrast, sulfur is only produced in significant quantities by the most massive stars, which are correspondingly rare (less than 1 % of the universe's stellar population). Consequently, hydrogen sulfide is considerably less abundant than water – hydrogen oxide. Possibly, without exception, life will have adapted to use what is most available: water. Thus, although hydrogen sulfide is likely to be present on all, or at least most, planets, water will be the dominant solvent, and life will most likely evolve to use it.

Water, too, is unique in its chemical properties. It is liquid, and hence a solvent, at modest temperatures. The nearest sibling molecules, including ammonia, are liquid at far lower temperatures than water – so low that the rates of the chemical reactions necessary for complex life are too slow for viability. Moreover, water molecules can give up and accept hydrogen ions with relative ease, putting lesser molecules, such as hydrogen sulfide, ammonia and hydrochloric acid, to shame. This simple chemical trick allows water, at modest acidity, to participate in the wide variety of chemical reactions that are essential for life. If the universe was wired differently and hydrogen sulfide or ammonia dominated, life might well be a rarity.

By functioning as both a solvent and a reactant, water is a fairly unique substance. Liquidity at modest temperatures also allows the other terrestrial molecules of life – amino acids, carbohydrates and nucleic acids, to name a few – to remain stable. Thus, despite the higher energy demands imposed on organisms that use water in photosynthesis, water is far superior to hydrogen sulfide or ammonia as a reactant and solvent.

The universe seems fairly well set up to use water for life. There may well be some unusual worlds in which hydrogen sulfide is dominant, but they are likely to be rare, to say the least: water is king. Given water's ubiquity, it would seem surprising if life on a red dwarf planet couldn't develop the necessary processes allowing it to steal hydrogen from this highly prevalent solvent. Given sufficient evolutionary time we can at least imagine that life will evolve the machinery necessary to split water and release hydrogen, even where the available energy is less than is found on Earth. We are reminded that absence of evidence does not equate to evidence of absence. A lack of such machinery on Earth certainly does not equate to an absence of such machinery elsewhere. Life is adept at surmounting challenges.

If plants arise that produce oxygen, life on a red dwarf world will require some unique adaptations. Research work carried out by Nancy Kiang (NASA Goddard Institute for Space Studies) and colleagues was presaged by that of James Kastings and later Martin Heath in the 1990s. Nancy Kaing and colleagues' work identified a number of challenges and possible solutions to these obstacles. These are discussed in the next section.

Physical Constraints Imposed on Photosynthesis by Red Dwarf Stars

All worlds that orbit within the stellar habitable zones of red dwarfs will be tidally locked to them. A fixed source of light in the sky imposes unique geographical constraints on vegetation cover. Plants located under the point of greatest illumination (the substellar point) will have to compete for the radiation coming from directly overhead. Shading of leaf tissue by any overlapping leaves will limit, or prevent, photosynthesis occurring in those tissues, *if* it is driven by visible light.

On Earth, terrestrial plants can turn their leaves towards the Sun like ships moving their sails in response to a changing breeze. However, the sun of a crimson world occupies a fixed location in the sky, so this form of shade avoidance will not work. Light and shade are permanent features, never moving from hour to hour; thus if plants are to compete for light they will have to physically move around shading obstacles. This could lead to some intriguing evolutionary arms races as plants battle for supremacy in their command of light.

As the distance from the sub-stellar point (SSP) increases, illumination will come from progressively shallower angles, leaving land plants the job of angling their stems and leaves so that the leaf surface is positioned at 90° to the incident light. Once again, shading by any geographical feature, as well as any intervening plant or animal, will prevent photosynthesis in these tissues. We could speculate that any advanced resident species could use the angle of plant leaves to calculate their latitude and confirm early on that they were living on a sphere. One might imagine a Proxima Centaurian using the angling of leaves of indigenous plants to determine the diameter of his or her home world, in much the same way as Eratosthenes used sticks in Alexandria and Syene (Aswan).

Furthermore, any geographical feature, such as a range of mountains, would block light to areas of the surface. The slow ebb and flow of orogeny and erosion would thus eliminate and then renew areas of land for colonization by any species of photosynthesizing organism. The slow movement of plates would also necessitate the slow but constant readjustment of plant life to changing light levels from generation to generation as their abodes drifted across the globe.

In response to a fixed point of light, land plants might even be physically mobile, perhaps slowly moving across the landscape in response to changes in their environment. Assuming plants are physically bound to the land like ours, the simplest coping strategy to geological change would be to produce wind-blown seeds or spores that would allow the passage of the organism's genetic heritage from one location to another. This is clearly not a disadvantage on Earth, where plants readily colonize new territories even where the source of seeds is a considerable distance away.

Earthly plants also colonize areas using animal hosts to transport seeds, either attaching themselves to the animal or by enticing the animal to consume their fruit. The latter strategy has the advantage that the seeds are then spread in fertilizing feces.

Finally, the dominantly red light behaves differently from blue light as it traverses the atmosphere of any habitable world. Red light scatters less efficiently off particles in the air than blue. This means that plants nearer the terminator will receive proportionately less visible light than they would otherwise do if the light were bluer. Thus, the amount of useful radiation plants would have at their disposal will fall off more quickly with distance from the SSP than would be found were the light source a star like our own.

However, the effect of a fixed sun is less stringent when photosynthesis is driven by infrared light compared with visible light. Unlike visible light, infrared radiation is emitted by every surface warmed by the parent star, as well as by molecules and particles in the atmosphere that are irradiated by the star. On a planet such as Earth, illuminated by a yellow star, the vast majority of the infrared radiation that permeates the atmosphere comes from the illuminated surface of the planet. This absorbs the predominantly visible light, and re-radiates it at longer, infrared wavelengths. Yes, there is infrared radiation from the Sun, but proportionately it constitutes a far smaller proportion of the total stellar output than is found from red dwarf stars.

Consequently, any world orbiting a red dwarf is thoroughly bathed in infrared radiation. It comes from the star, the planetary surface and the atmosphere. Any photosynthesizing organism that uses infrared radiation can obtain energy from nearly every direction; shading is thus of little consequence as the plant is irradiated from all sides. Therefore, in principle, if plants are adapted to absorb sufficient infrared radiation, then they could survive even when shaded from direct sunlight (Fig. 8.6).

In water, the effect of scattering and refraction of visible light is even more pronounced than it is in air. To survive aquatic plants and photosynthesizing single-celled organisms will need to grow nearer the top of the water column than those on Earth. Once the rate of photosynthetic production falls to 1 % of that found at the surface of our planet, production of food molecules

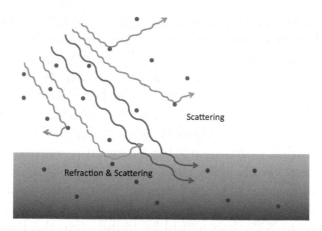

FIG. 8.6 The scattering of light by particles in air and water. Blue light is scattered far more effectively than red light, as the wavelength of blue light most resembles the size of the particles. Conversely, waves of red light essentially weave around these. The effect explains why our sky is blue and sunsets are red. In a column of water light is also effectively refracted so that it quickly becomes dark as you descend. This limits the depth at which photosynthesis can occur

by photosynthesis matches that consumed by respiration. At this point any plant or microbe attempting to make a living from its sunlight could not make enough food to sustain itself. The depth at which this transition is reached is called photic window. On Earth, this is typically up to 200 m in depth in open ocean. However, nearer coastal areas, where run-off from adjacent land brings sediments or nutrients for microbial growth, this can be less than 15 m in depth.

Unfortunately, light in the green and yellow parts of the spectrum are the most penetrating, while useful red and blue light are effectively blocked by this murk. Green and yellow light are exactly the regions of the visible spectrum used least well by terrestrial plants. Therefore, they are of no real benefit for any plants or photosynthesizing microbes found on Earth. Infrared radiation is even less penetrating in water. Therefore, microorganisms adapted to the sole use of this type of radiation are likely to be restricted to the very top of the water column, or to adjacent land.

There is an interesting point here. Many people will be aware that 25 % of Earth's oxygen production comes from the Amazonian rainforest. Few, however, are aware that almost all of this output is reabsorbed by the forest to power respiration. Instead, Earth gets most of its oxygen from the oceans – from algae and bacteria that ride the currents. On the planet of a red dwarf, an ocean that is too turbid for photosynthetic life will give rise to a planet that is effectively free from oxygen. Moreover, photosynthesis evolved first in the oceans or other bodies of water on Earth. Conceivably, this could be impossible on a crimson world if photosynthesis is restricted or altogether absent in the oceans of these worlds. This might be one more obstacle to the rise of photosynthesis on such worlds.

Once again, although the depth of the photic zone imposes restrictions on the habitability of the red dwarf world, it does not preclude it. The depths of our oceans are dark, and no plant life is present. Yet the oceans are hardly barren, even at considerable depth. The abyssal plains that stretch endlessly across the floors of our oceans are populated by scavengers, decomposers and microbes that derive energy from geological processes, such as volcanic activity. A deep ocean area opaque to light does not prevent the spread of life.

The Effect of Planetary Climate

The climate of the red dwarf could also impose restrictions on life if tidally locked planets experience permanent storm features over areas of land or sea. In equatorial regions on Earth daily storms arise as a result of strong heating of moist air by the Sun. Convection generates dense cumulonimbus storm clouds that bring evening rains. It is conceivable that the formation of a permanently cloudy, wet area over the region of strongest heating (the SSP) could limit the availability of light. Although this region would be expected to experience the best lighting and heating in a cloudless world, in a more reasonable planet with an adequate hydrosphere, cloud cover could block sufficient light to preclude active photosynthesis, at least periodically.

Of course, this is an extreme scenario. On Earth clouds come and go in response to the ebb and flow of convection, the movement of air and the finite lifetime for most storm systems. On a tidally locked world these same constraints are likely to apply, assuming that some form of slow-moving and self-sustaining storm system does not form. This will require further computer modeling and future observation, and even the best guesstimates from modeling work will be heavily affected by the nuances of planetary topography.

The Temperature Limits of Life

Life is highly adaptable on our planet. Organisms are found growing in ice at temperatures of –5 °C or less. At the other extreme, some species of bacteria can survive temperatures in excess of 100 °C, the boiling point of water at atmospheric pressure. The limiting factor, as far as we are aware, is the requirement for liquid water. This is hardly surprising as life depends on the intricate and unique properties of this molecule.

Water is a solvent. Thus, where water remains liquid, we can probably find living organisms of one sort or another. At temperatures below the expected freezing point of water (0 °C), liquidity is possible as long as there are other chemicals present, dissolved in its mass. These depress the melting point so that either water stays in its entirety in its liquid form or partial melting concentrates solutes in narrow pockets or veins of liquid, woven into a frozen, purer water matrix. In these environments living bacteria and algae are found growing on rocks, within rocks or within or under ice in places such as Antarctica or high in the Andes.

More complex animals and plants require more restrictive climes. Heftier organisms, with large numbers of interdependent cells, require temperatures that allow chemical reactions to persist at more substantial rates necessary for their viability. A mammal will not prosper if its core temperature falls much below 33 °C. Many mammals will either migrate or hibernate. While hibernating they turn down their thermostats and go into an extended sleep mode. They survive, but they are unable to reproduce in this state. Clearly an animal will not reproduce while it is asleep. Such an animal may well survive, but it cannot prosper.

At the opposite extreme most organisms suffer when temperatures exceed 40–45 °C. Some animals have adaptations that allow them to tolerate these temperatures for part of the day, and often this necessitates efficient management of body water content as well as thermostasis – management of temperature. Most complex animals will alter their behavior to limit exposure to high temperatures, often burrowing underground until the evening brings cooler conditions. However, on a tidally locked planet there are no evenings. Migration to cooler latitudes or hiding behind rocks or other structures may offer protection, but it is unlikely that they could do this on a continuous basis unless suitable prey, or their predators, were similarly obliging. Simply avoiding the hottest part of the planet's surface would seem to be the logical strategy.

On Earth, higher plants are restricted to temperatures in the range –5–45 °C. At the high end of this range photosynthesis is inhibited by disturbances in the functions of the enzymes and components needed to capture carbon dioxide. Grasses are somewhat hardier than their broad-leafed siblings and will carry out photosynthesis at temperatures as high as 60 °C, although they clearly suffer as a result. Again, like their animal cousins, they survive but are hardly prospering.

Photosynthesizing bacteria seem to do much better and can function adequately at temperatures as high as 75 °C. Thus, wherever we look, the most abundant life-forms are likely to resemble our highly adaptable, single-celled bacteria.

Finally, most Earthly plants are unable to photosynthesize if annual temps are less than 10 °C for more than 1 month a year. Therefore, on any tidally locked red dwarf world, we will be looking for life on planets where substantial areas of the surface experience temperatures in the range of 10–45 °C. Elsewhere bacteria-like life forms may be the norm. However, let's not presuppose what marvels evolution may provide in response to these challenges. More often than not there are more exceptions than there are rules.

The Effect of Star Spots

A final complication to life under a red sun is the influence of star spots. Here on Earth sunspots have little immediate impact on terrestrial weather. They cover a small fraction of the Sun's surface even at solar maximum, when numbers are highest. Examination of young red dwarfs shows that in principle large numbers of dark spots would reduce the total amount of radiation by as much as 40 % on the surface. There would be a substantial cooling on any planet orbiting a particularly spotty star in this circumstance. If the planet had a marginal climate, this could reduce temperatures by over 20°, allowing substantial freezing of its surface.

This may seem troublesome, yet the effects wouldn't last long, perhaps no longer than one of our winters. Moreover, star spots, much like acne, are largely the preserve of youth. By a billion years or so of age most red dwarfs rotate more slowly, having braked their rotation against the interstellar magnetic field. In turn, this weakens their magnetic field. With a weaker magnetic field, such mature stars have substantially less pock-marked faces; consequently the impact on planetary habitability is measurably less.

Furthermore, we know from observations of terrestrial climate sunspots, or, more precisely, the position the Sun is in that its cycle appears to have some, albeit a rather weak, effect on climate. Far from being the harbingers of frigidity sunspots appear instead to be the harbingers of milder conditions. On Earth times of high sunspot activity are associated with warmer summers in the northern hemisphere, although there are many exceptions to this generic rule. Moreover, there are hints of longer term variability linked to a lack of sunspots, rather than their excess. The Little Ice Age in seventeenth-century Europe and North America was associated with a prolonged absence of sunspots.

Although not yet substantiated, higher levels of sunspot activity may be associated with reductions in cloud cover, causing greater irradiation of the surface. Alternatively, it is also proposed that high sunspot activity can lead to alterations in the circulation of the middle atmosphere, which then feeds back to conditions at the surface, causing heating. Our understanding of the meteorological impact of sunspots is rudimentary at best. However, overall, the observed effects are substantially less than 1 °C and consequently not likely to cause substantial harm to any biosphere.

Observation of our Earthly climate demonstrates the impact of increasing levels of carbon dioxide, but there are more interesting and counterintuitive consequences of our anthropogenic warming of the atmosphere. In Europe the current succession of cold, miserable summers may be a consequence of the overall warming trend.

Now, this may well sound like an oxymoron – warming causes cooling. However, there is an interesting and perhaps unexpected chain of logic here. Let's follow it.

In the north, ice is melting at an unprecedented rate, and the Arctic could well be ice free in summer by 2050. Greenland's cap is also melting and thinning at a frightening rate. So far so good – warming causes ice to melt. But this is where it gets more interesting. Less ice means more sunlight is absorbed into the oceans, and bedrock causing rapid warming, particularly in the summer. This warming generates prolonged, warm-cored high pressure areas over the high latitudes. Normally, the summer months are dominated by cold-cored low or high pressure areas at these latitudes. However, warm-cored high pressure areas block the normal easterly progression of low pressure zones, with their incumbent cloud and rain. These are spawned and driven from west to east by a narrow band of winds called the jet stream. Instead of a clean west-to-east flow the northern block diverts the rain-bearing lows to the south.

However, land is very unevenly distributed in the northern hemisphere. Rather than a smooth shift southwards, the alignment of the continents drives rain-bearing low pressure areas south across the Atlantic into western Europe. Conversely, over North America, the jet stream bulges northward, bringing prolonged drought. Meanwhile, over the Arctic, cloud-free high pressure areas generate more melting of the ice and so the pattern of melting and wind-flow change is accentuated.

There has been an interesting trend in European climate in the last decade. Winters have been colder and much drier, although for the UK the number isn't yet exceptional. Above and beyond this, the last six summers have been wet as a result of the jet being forced south. The years 2011–2012 were a notable case in point. After a drought lasting from late August to April, the fields of the UK turned to dust. Mid-way through April all changed as the jet stream launched a succession of rain-bearing low pressures that dragged on through to September 2012. June and July brought exceptional rain, exceeding even the deluge wrought in 2007. The pattern of British weather now appears to be dry and often mild autumns; cold, dry winters; and warm, dry springs. Just as our farmers come to terms with this, the summers drown the crops.

Meanwhile, parts of North America labor under prolonged drought. The jet stream is nowhere in sight.

These somewhat counterintuitive effects of anthropogenic global warming should inform our thoughts of alien worlds, and vice versa. Climate is a complex beast that will take a long time to understand, never mind tame. Thus, when we turn our thoughts to distant worlds, orbiting unconventional suns, we should be wary of applying our often simplistic anthropocentric views. We are often better at discussing our models than we are at creating them. The effect of star spots on the climate and hence life will need a lot more thought before we are able to describe its effects in detail.

Star spots are also associated with violent flaring. On our relatively aged Sun they cause rapid increases in X-ray and ultraviolet output but are relatively infrequent. Aside from impacting our satellite and electronic systems, flares have little effect on life on Earth. However, flares on red dwarfs are substantially more energetic, and because the underlying output of the star is so low, the flare generates a proportionately greater increase in stellar radiation.

Flares on a population of red dwarfs – the so-called UV Ceti stars – are also frequent, occurring as often as every 20 min. Any living organisms would have to adapt to wide variations in the amount of ultraviolet and blue light emitted by their host star. Most X-rays and ultraviolet light would be blocked by the planet's atmosphere, but most of the blue light would make it through. Again, Earthly plants respond to the stress of rapid increases in visible light by increasing their production of purple pigments called anthocyanins. These can turn leaves dark brown, purple or even black. The plant survives, but the yield of photosynthesis goes down as less light makes it to the photosynthetic pigments. Nancy Kaing and others suggest that the yield of usable substances from photosynthesis of plants on red dwarf worlds may only be a fraction of that found on Earth, perhaps as low as 4 %. This would not preclude active photosynthesis but would restrict plant growth. Yet, with perpetual daylight on most habitable red star planets, this is unlikely to prevent the growth of higher plants.

We should also remember that a plant photosynthesizing on a red dwarf world may already have black leaves. The unconventional color would be a response to the lower availability of visible light and the preponderance of infrared radiation. Thus the occasional flare may at worst be utterly harmless or at best an additional, useful source of high energy radiation to supplement that available for photosynthesis.

In the longer term frequent flaring could lead to erosion of the planet's atmosphere. Violent increases in the stellar wind, and increases in heating of the upper atmosphere, lead to its expansion and erosion. The effect will be small each time, but over billions of years could lead to loss of a substantial mass of atmosphere. This is likely to be by far the biggest long-term hazard to life on a red dwarf world (Chap. 10). A planetary magnetic field will offer some protection by diverting the worst excesses of the stellar wind. Yet, planetary magnetic fields require a circulating liquid iron core, which in turn is dependent on reasonably fast planetary rotation and core convection.

However, tidally locked planets rotate slowly, once per orbit, and thus may be unable to generate magnetic fields strong enough to protect their world from the atmospheric erosion. Moreover, plate tectonics may be a preserve of planetary youth. Older planets, on which plate tectonics has ceased, may also lack protective magnetic fields. Plate tectonics brings cold material down into the lower mantle (Chaps. 4 and 5). This in turn generates sufficient temperature contrast within the liquid core, allowing it to convect. Super-terran worlds may thus start out able to defend themselves from the ingress of their parent star's wind. However, if their crust congeals into a thick lid (Chap. 5) or if they are tidally locked to their host star, planetary magnetism may fail early on and the atmosphere may become fatally eroded. A super-terran may not be so super after all.

Thus there are a number of important constraints on the development of photosynthesis on planets orbiting these red stars. These constraints are unique to these worlds, where the planet is tidally locked. In the unlikely scenario that the processes of mutation and natural selection were unable to compensate for these, life might be restricted to planets orbiting more massive red dwarfs. Here, the stellar habitable zone would extend outside the tidal-locking radius, and planets would at least rotate slowly relative to its star.

This may be true for Gliese 581d, where the four-planet solution for the radial velocity data suggests that Gliese 581d lies at least part of the time outside the tidal-locking radius (Chap. 9). Gliese 581d may thus rotate slowly, perhaps two times per its 160 h orbit. This would give it a day length marginally less than four of our days. However, for planets orbiting the most abundant, less substantial stars tidal locking would be a certainty. A planet orbiting a star with one tenth the mass of the Sun takes approximately two Earth-days to do so and cannot avoid tidal locking. Therefore, the physical constraints of having a star occupy a fixed location in the sky and a low energy power source could conspire to restrict photosynthesis. Once again, many of these issues seem rather whimsical, given our observation that life tends to surmount challenges imposed upon it.

Conclusions

The power of natural selection derives from large numbers. A liter of water holds nearly a trillion trillion molecules. A cell holds trillions of dissolved substances, and an ocean in which life arises trillions of times more. A 100 million years with quintillions of molecules offers almost boundless possibilities for chemistry to explore.

Within a teaspoon of bacteria a 100 billion cells mutate, exploring the spaces afforded by their environment. Evolution by natural selection is as much a certainty as the Sun rising in the morning. The chemistry of a red dwarf world may well inhabit a different space, but the possibilities are still endless. Most likely it will be our imaginations that fail to rise to the challenge, not the chemistry underlying evolutionary development.

Red dwarfs generate some unique challenges for living organisms, both in terms of their initial development and their subsequent evolution. Many of these are not obvious at first glance, for example the radiation produced by these stars. The constraints imposed by having a fixed light source in the sky and the low energy of the available radiation could make life's journey difficult, but almost certainly not impossible. The limited capacity of infrared radiation to permeate deep into water may be a problem for life, as may be the absorption of infrared radiation by compounds in the atmosphere. Failure of the geodynamo might cause the planet to lose its atmosphere in much the same way as Mars has, or a super-terran's geology may flood the biosphere with so much hydrogen sulfide that oxygenic photosynthesis is prohibited. Yet, if life responds to infrared radiation and uses it to drive photosynthesis, many of the constraints on photosynthesis by plants vanish. Infrared radiation can, in effect, go around corners and reach the parts of a planet's surface that visible light cannot.

Consequently, the first living world, outside our Solar System, is likely to be found orbiting a nearby red dwarf. There are simply so many of these star systems that it seems highly improbable that life would not find a way. What a fantastic discovery this would be. With the galaxy filled to the bursting point with 150 billion or more red dwarfs, and perhaps all of these stars hosting planetary systems, do the math. A quadrillion chemical possibilities on a 100 billion worlds. Advanced life based on oxygenic photosynthesis seems likely to the point of certainty.

9. Gliese 581 d: The First Potentially Habitable Water-World Discovered?

Introduction

Skulking in the shadows of nearby brighter stars lies the 581st member of the Wilhelm Gliese's catalog of nearby dim stars. The catalog contains nearly 1,000 stars organized according to their position and stellar properties. Gliese 581 is a low mass, red dwarf star weighing in at a little under one third the mass of our Sun with a surface temperature of 3,200 °C. This little, unassuming star lies 20.3 light years from Earth. It's a diminutive body of light and color, and its low content of metals suggests an age of 7 billion years or more. When the Sun formed, Gleise 581 had already settled down and was, by our standards, a mature star. Any instability in either the orbits of its progeny worlds or its surface would long since have ceased. Its planetary retinue circumnavigate their parent star in tight, nearly circular paths.

In the middle part of the first decade of this millennium the spectrum of the star came under the scrutiny of a number of astronomers. In 2005 the first two planets, b and c, were found by Paul Butler and Steven Vogt. By 2009 a total of four planets had been confirmed through subsequent investigations by Geoffrey Marcy and Stéphane Udry. One of these, Udry's Planet d, looked as though it nestled near or just within the habitable zone of its star. Gliese 581d had a mass roughly six times that of Earth and claimed the title as the first potentially habitable planet to be discovered. Shortly thereafter, another two worlds were proposed by Steven Vogt's team, Gliese 581 f and g. Its modest mass suggested that Gliese 581d might be a aquaplanet – a world with a surface wholly or mostly covered in a deep ocean of liquid water.

D.S. Stevenson, *Under a Crimson Sun: Prospects for Life in a Red Dwarf System*, Astronomers' Universe, DOI 10.1007/978-1-4614-8133-1_9,
© Springer Science+Business Media New York 2013

Gliese 581 has a tight little system of worlds. All of the orbits of the confirmed planets lie within the orbit of Mercury were we somehow to transplant its worlds to our Solar System. This arrangement may seem alien, with four sizable planets squeezed within the orbit of tiny Mercury. However, observations of alien worlds by the Kepler mission show that this is by no means unusual. Many of the stellar systems probed by Kepler show a retinue of worlds confined to orbits deep within the gravitational wells of their parent stars. Indeed, red dwarf stars seem to prefer to keep their worlds close, snuggling each tightly to their small fire.

One has to remember that Gliese 581 is a small star, only three times the diameter of Jupiter. Consequently, the Gleise 581 planetary system resembles the Jovian Galilean satellite system rather than our Solar System as a whole. The planetary orbits are scaled down in relation to the shrunken diameter of the star itself. In large part this is a consequence of the formation process for these systems. The protoplanetary discs from which they form are much smaller and less massive (Chap. 2) than that which formed our Solar System. Rocky or Neptunian worlds appear to dominate these systems, and Jupiter-sized worlds are rare. None are as expansive as our Solar System in scale, reflecting a birthing chamber that is considerably smaller than that of the developing Sun.

This is not to say some Sun-like stars don't mimic the layout of red dwarf systems. Some of Kepler's discoveries have similarly constrained orbits around much more massive, Sun-like stars. Kepler 11 is home to six tightly nested planets. Five reside well within the equivalent orbit of Mercury in our system, while the sixth would lie within the orbit of Venus. And these are big worlds – super-terrans or Neptunian orbs.

Therefore, we must cast aside our preconceptions regarding the nature of other stellar systems. Our Solar System, although not unique, is by no means the model by which all other planetary systems abide; there is great diversity. Indeed, Carl Sagan's *Cosmos* illustrated this as early as 1981. In one chapter Sagan presented an array of possible planetary systems based on the computer models of the day. The Solar System merely displayed one of many modes of planetary arrangement. Many of the others had what appeared to be very alien constructions, but observations nonetheless confirm their existence.

Since red dwarfs are by far the most common stellar factories in the universe, it would seem probable that dense planetary systems such as Gliese 581 are the norm, rather than the exception. It will be the worlds of red dwarf stars that hold the greatest retinue of living organisms, not those of more extravagant stars such as our Sun.

Gliese 581g: Now You See It, Now You Don't

Gliese 581 has proved to be a frustrating write. In 2010 Steven Vogt's team presented evidence of six planets orbiting the central star – all in circular orbits, much like the Kepler 11 system. Data from radial velocity measurements was used to support this assertion. Here, slight perturbations in the motion of the central star were caused by the gravitational tugs of the circumnavigating worlds. Planets b and c were discovered by 2007 by Steven Vogt and Paul Butler using measurements taken by Keck's HIRES and the ESO's HARPS instruments and were regarded as thoroughly solid finds. Planets d and e followed a short while later.

Despite the perturbations of each planetary orbit by the neighboring bodies in the system, astronomers felt sure that all four planets existed. Complex statistical analysis is required to separate the gravitational effects of each individual planet on the host star, as well as more complex overtones that each planet has on one another. Together these gravitational tugs can make the investigation of these frequently compact planetary systems challenging, to say the least. That said, no one doubted the existence of Planets b, c, d and e.

However, in 2009, Vogt published a further analysis of the HARPS and HIRES data, claiming to have identified the signal of two further planetary bodies, f and g (Chap. 1). The inherent risk with this approach – a form of meta-analysis – is that when combining data sets one also combines instrumental noise. Meta-analysis is frequently used in a variety of settings in biological and psychological research, where the strength of individual studies may be weak, but cumulatively the data is strong. The intention is to get the best out of all included studies.

However, the potential for summing instrumental noise is that this may create unintentional problems – false signals that appear simply to be partially obscured by noise, rather than noise itself. From his analysis Planet f lay squarely outside the realms of habitability. If that was it, then nothing else would have mattered. However, it wasn't. According to Vogt's calculations Planet g appeared to lie in the middle of Gliese 581's habitable zone; thus the stakes were raised considerably.

In December 2009, a research group led by Rene Andrae of the Max Planck Institute for Astronomy in Heidelberg, Germany, submitted a paper asserting that Steven Vogt's team had incorrectly assumed that the six Gliese 581 planets had circular orbits. The requirement for two additional worlds was eliminated when eccentric orbits were substituted for the circular ones of Vogt. The perceived additional signals were merely background noise in the HAPRS and HIRES radial velocity spectra of Gliese 581: planets f and g weren't there.

In September 2011, Thierry Forveille and colleagues (Observatoire de Grenoble) went further, adding another 121 radial velocity measurements from HARPS to those already found by HARPS prior to 2009. They asserted that Planets f and g were unnecessary to explain the radial velocity measurements for the system. Occam's razor sliced away the need for the two additional worlds.

In January 2012 a final nail appeared to be hammered into the coffin of Planet g. Philip Gregory (University of British Columbia) submitted a paper in which he re-analyzed the available HARPS and HIRES data that Vogt's team had used. Gregory used Bayesian statistical techniques and once again found a best fit for four worlds in eccentric orbits, rather than six in circular ones. Planets f and g seemed to melt away (Fig. 9.1).

The systems implied by these data indicated a modest level of eccentricity in the orbits of the four planets, notably Gliese 581 e and d. However, would such a system remain stable over its suggested 7 billion year lifetime? More pertinent, was Gliese 581g – the most habitable exoplanet to date – a dead world? To both questions Steven Vogt thought not, and vigorously pursued further analysis of the available data.

The major concern expressed by Vogt was the stability of the suggested four-planet solution. Vogt and colleagues ran computer

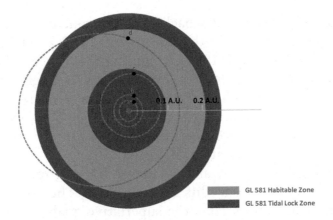

FIG. 9.1 A reiteration of Fig. 6.1. The four-planet solution for the Gliese 581 planetary system, based on the HARPS data. Planets *b, c* and *e* are all tidally locked to their parent star, while *d* may be tidally locked or experience pseudosynchronous rotation, with days lasting almost a full planetary year

simulations for each suggested four-planet solution and found all to be unstable over only a few tens of thousands of years. If this interpretation was correct clearly something was amiss with the four-planet solution. Vogt then reanalyzed the complete HARPS data set and found an all-circular four-planet solution worked reasonably well. So, where if anywhere was Planet g, and how could an apparent need for orbital eccentricity in the four-planet model be reconciled with their apparent instability?

The modeled data set used by Forveille omitted some of the original data from its analysis. Vogt included this and with a reasonable level of significance, Planet g appeared to resurface. The apparent eccentricity falls away once a fifth planet is included with an orbit of roughly 35 days. Thus Planet g reappears as a means of securing circularity and stability, rather than because of an overt signal in the data. The reemergence of Planet g is thus an indirect conclusion, but one that seems reasonably robust. Vogt is now busy collecting and analyzing more data from HARPS as it comes in. Clearly this is essential. The data is far from conclusive at present. No doubt there will be further twists in this already complex tale.

Vogt's interpretation of the data appears thorough, but one suspects that his use of statistical analysis will be challenged again.

With the ball served neatly back to the opponent's court at present, we can conclude that the insertion of the 2.2–3.1 Earth-mass planet Gliese 581g, at 0.13 A.U., appears to work as a best interpretation of the available data. Interestingly, if Vogt's interpretation is correct, planet Gliese 581d will also move back into a more "habitable" orbit around its star, at 0.21 A.U. This would make Gliese 581 home to two potentially habitable worlds – another first for the planetary system.

Although the to-ing and fro-ing of Planet g might suggest a muddled state of affairs, this is misleading. The rise and fall of data in this story illustrates the superlative qualities of hard science. Assertions based on incomplete evidence were swept this way and that, but in the end (if this is indeed the end) data drives conclusions. It is this property that elevates science above many other facets of modern life. Reappraisal of work exposes weakness and either strengthens or eliminates it, leaving a much healthier body behind.

As the number of detected worlds expands, at seemingly geometric rates, we must place our enthusiasm and pride far behind the colder, prosaic reality of hard data. What is always disappointing is the insidious infiltration of personality. A battle has raged over the existence of Gliese 581g – one that may well not be over. On more than one occasion public acrimony has erupted over the presence or absence of this potentially habitable world. A cool head is always preferred. Planet g may yet evaporate once more, but at the time of writing the data from HARPS seems to support its existence with fairly high confidence.

Until the case is finally closed, Planet g will be "ignored" in further discussion. The other habitable world – planet 581d – will form the centerpiece of this chapter. If Gliese 581g is confirmed, it easily outranks its siblings in terms of habitability. Its low mass – perhaps only a little over twice that of Earth – could mean that it is a rocky world with a substantially dry surface – as well as oceans of liquid water. With a lower mass than Gliese 667Cc (Chap. 10), this may be all the more true. However, with a tide of information that seems to ebb and flow from month to month, it will be worth hanging fire on further discussion. What is important to stress is that with a similar stellar composition and perhaps age to 667Cc, any consideration of 667Cc will transpose effectively to Gliese 581g.

The uncertainty over age – and the obvious lack of direct information about the nature of the bodies – means that the following discussion of planetary habitability must be considered hypothetical. However, it does follow the rules discussed in the earlier chapters. These are defined by the geophysics and geochemistry of planets, the chemistry of life (as we know it) and the synchronous evolution of both. Given its mass and likely composition, Gliese 581d forms the centerpiece for discussion of aquaplanets, while Gliese 667Cc forms the focus of our discussion of dry, Earth-like worlds (Chap. 10).

Gliese 581d: An Ancient Water World?

Despite retaining its title as the first potentially habitable world to be discovered, the aging Gliese 581d may not be the nirvana it was first thought. Without giving too much away some uncomfortable truths about this world need to be faced. These might relegate Planet d to the wilderness of inhospitability.

Before plowing into what could well be bad news, we review the star system and put the discovery of Gliese 581d in context. The red dwarf Gliese 581 has a mass of a little over three tenths that of the Sun, with a radius somewhat less than one third that of our star. The star has a relatively low proportion of heavy elements compared to our Sun – roughly one third – implying that it is old, perhaps as much as 11 billion years old. Whatever its true age it is likely to exceed 7 billion years, making it more than half the age of the Milky Way Galaxy (Fig. 9.2).

Within the astronomical community there is much wrangling over the link between the abundance of metals (elements heavier than helium) and the ability to form planets. The book *Rare Earth* by Peter Ward and Donald Brownlee (Springer, 2000) went so far as to claim the existence of a galactic habitable zone that was bounded on the outside by stars with an insufficient abundance of metals needed to construct planetary bodies. Given the scarcity of evidence at the time of publication, this seemed like a very bold, if somewhat foolhardy proposition.

Events soon confirmed this suspicion with the discovery of a Jupiter-mass world in orbit around an uncomfortable pairing of a

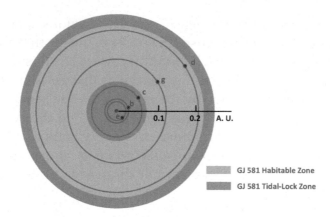

FIG. 9.2 The revised 2012 orbital model of Steven Vogt et al., based on his extended analysis of HARP and HIRES data. Planet g makes a reappearance, near the middle of Gliese 581's habitable zone. With a minimal mass of just 2.2 Earths, GJ 581g returned to top spot as the most habitable exoplanet found at the time of writing. The planets and star are not drawn to scale

white dwarf and millisecond pulsar (Chap. 1). This tortured world was hiding within the denizens of the metal-poor Population II globular cluster M4. Thus although there seems to be a preponderance of planets in stars with super-solar metal abundances, there are plenty of other worlds linked to more impoverished stars. Thus claims that metal-poor stars simply won't host planets are unfounded and based more on limited evidence or poor supposition than on hard data.

With Gliese 581 weighing in at such a low mass, this M-dwarf releases only 0.2 % of the visible radiation of our Sun. However, with a spectral peak in the infrared, the total radiation released is more like 1.3 % that of the Sun. The total output seems trivial beside our Sun's, but the distinction between visible output and total output is important. More than any other wavelength the abundance of infrared light will determine the temperature on the surface of any planet orbiting such a low mass star. Stellar output is grossly underestimated for both the least and most massive stars. The least massive stars have radiation dominated by the infrared, while the most massive have an output dominated by the ultraviolet. Visible output does clearly vary with mass, but because most telescopic surveys concentrate on visible light, they understate the energy from the stellar extremes. Determination of habitability must, therefore, take into account the full stellar energy budget, not just what we can see with our limited eyes.

What of the planets themselves? Planets b, c, d and e were all found using HARPS based at the European Southern Observatory in Chile (Chap. 1). Planet e nuzzles in closest to its parent. At the time of its discovery it held the title as the smallest identified world outside the Solar System – another first for the Gliese 581 system. Orbiting less than 2 million kilometers from the photosphere of Gliese 581, Planet e is a 1.94–3.10 Earth-mass sizzler. Liquid water is impossible, and the surface is likely baked to temperatures in excess of 700 °C.

Whether Planet e holds onto an atmosphere is problematic, but given its relatively high mass it is not impossible. However, it will not be an atmosphere anyone would want to breathe. Planet e whizzes around its parent star in 3 days, always presenting the same face to its host star. The view from the starlit surface would be impressive. Red Gliese 581 would occupy 70 % of the sky, with red arches of gas frequently pouring off its surface towards the broiled world. In the absence of an atmosphere the permanently dark night side would freeze at temperatures of less than –180 °C. If Planet e holds onto a suitably dense atmosphere, perhaps like that of Venus, gases will be forced to rush headlong from sunlit to dark sides in a desperate battle to even out the temperature extremes. In such a Venusian world, even the night side would be unpleasantly hot.

Moving further out, we encounter Planet b. This 15–30 Earth-mass world may well resemble Neptune, but would be far hotter. Once more temperatures could exceed 400 °C, particularly if the atmosphere retains much of the incident radiation through the action of greenhouse gases such as water vapor, carbon dioxide or methane. Planet b is also tidally locked to its Sun, taking 5 days to swing around its host. Given the strong heating and probable thick atmosphere, the weather on such a world is likely to be wild, as planetary winds seek to even out the temperature differences from starlit to night sides.

On the inner edge of Gliese 581's habitable zone lies Planet c, orbiting 11 million kilometers from its star. Weighing in at five to ten times the mass of Earth this world might be a diminutive version of Neptune or something altogether different, a planet dominated by a hot, thick ocean of water and a deep, overwhelming body of cloud. Planet c orbits its host star every 13 days, and, like Planets b and e, is tidally locked to the star.

Given that red dwarfs are more luminous before birth than they are once they reach the main sequence, Gliese 581c was most likely superheated at birth. This contrasts with Venus, which may only have achieved its present greenhouse state in the last billion years or so, as the Sun slowly brightened. With such strong, early heating, only a planet with a very rich supply of volatile gases would have been able to form oceans at all. Perhaps it is more likely that the planet was born dry with any water confined either to a dense, hot atmosphere or compressed into a layer of hot ices at great depths. Alternatively, this world was dominated by rock rather than the gas and liquid of Neptune. In such a scenario oceans would most likely never have formed.

Greenhouse warming would most likely have raised the temperature of Planet c to the point at which a runaway greenhouse effect occurred. Consequently, after a few hundred million years, any water will have most likely been split into its components, hydrogen and oxygen, with the former largely if not completely lost to space. On such a world the presence of any liquid water oceans will be restricted, held only in place if the planet has a very thick atmosphere than can pressure-cook the ocean below and prevent its total evaporation. It is not likely to be habitable except perhaps for the most extreme extremophile.

Finally, as we move outwards from Gliese 581, we reach Planet d, in a far more distant orbit. Planet d was discovered in 2007 by Steven Udry and co-workers and takes nearly 67 days (1,600 h) to orbit its red dwarf. Estimates of the planet's mass range from 5 to 6 Earth masses to as high as 10 times that of Earth. Thus Planet d could either be a super-terran or a small Neptune-like world. As Steven Udry commented, Gliese 581d may be the first discovered water-world, a planet bathed in a deep ocean of water. If Planet d is a water-world, the kinds of geological and atmospheric processes discussed in the earlier chapters will need closer examination and modification. Such a wealth of volatile material would alter the processes occurring in the planetary mantle and affect the evolution of gases such as carbon dioxide in the atmosphere. A planet-wide ocean would also likely restrict the types of life that could evolve there. These implications will be discussed in detail later in this chapter.

If there are only four planets in the Gliese 581 system Planet d will have a fairly eccentric orbit, stretching from 21 to 49 million kilometers from its star. The elongated orbit would take Planet d in and out of both the star's habitable and tidal locking zones, making its climate more unstable and its day length less than its year. However, if Planet g is confirmed, then d will orbit the central star at a distance of around 31 million kilometers, thoroughly locking one face to its source of heat and light. Without confirmation of Planet g's existence the more eccentric orbital option is considered.

The Gliese 581 system has a quite remarkable retinue of worlds. As was mentioned Planet e holds the record for the smallest planet found to date, and d for the first discovered potentially habitable one. The discovery of Planet d was important in regard to many scientists' view of red dwarf systems. Until more recent HARPS data came in, many scientists were of the opinion that red dwarfs were problematic hosts for habitable worlds. Not only were their habitable zones small, but their worlds within the habitable zone would be tidally locked. Astrobiologists began to look once more to the universe's more massive K- and G-class stars. These more massive stars host larger habitable zones that increase the chance that a planet will fall within it. Gliese 581 rekindled the interest of and re-focused the minds of astrobiologists.

Orbital Migration and the Composition of Gliese 581d

The general processes through which planets develop are reasonably well understood. Using what we already know we can infer the processes through which the Gliese 581 system was born, perhaps 7 billion years ago. Coalescing from a disc of gas and dust, the majority of the material settles into a contracting hot ball of debris called a protostar. Around 99 % of the material ends up in the protostar while the remaining 1 % forms a rotating disc of material around the developing star.

A star with a mass as limited as Gliese 581 would take hundreds of millions of years to condense from its natal sac. During this time important transitions were taking place within the surrounding disc of material. Some of these processes are likely common to all developing star and planetary systems, while a few are perhaps unique to red dwarfs.

The common processes involve the coagulation of planets from the debris disc and the orbital migration of these developing worlds as they grow in size. Uniquely, red dwarf stars become dimmer as they develop. A star like the Sun is very bright in its earliest stages of formation. However, soon after collecting the greatest fraction of its mass, the luminosity reaches a level that remains relatively constant as the bulk of the planets form. A red dwarf also begins life as a luminous ball of gas and dust. But as it shrinks in size its luminosity steadily decreases until hydrogen is ignited in its core. During this phase the distance from the protostar at which volatile gases like water can condense moves inwards in response to the diminishing heat. This process is important. During the greatest proportion of time that any surrounding terrestrial worlds are developing, they will most likely be bone-dry – baked under intense radiation from the contracting central star. Any volatile materials must be delivered after the main phase of planet formation is completed.

Therefore, if the Gliesian planets were likely all born dry, how could they obtain the volatile ingredients necessary for life?

We tend to assume that planets form in nice, tidy orbits and essentially stay as they are once formed. However, messages from our own Asteroid Belt paint a very different picture of planetary formation and evolution. Moreover, the unexpected detection of the first Jupiter-mass world in a tight and uncomfortably hot orbit around 51 Peg rang alarm bells in astronomical circles.

Jupiter, and presumably all other gas giants, were expected to form several astronomical units from their parental stars. Here, temperatures would be sufficiently low to allow gases to condense as icy grains. These would then rapidly assemble into larger bodies. An imaginary line can be drawn around the star beyond which ices condense. This is called the snow line. For stars like the Sun the snow line lies at about 4 astronomical units (A.U.). During the initial contraction smaller, red dwarf stars have a snow line nearer 2–3 A.U. Unlike Sun-like stars, this line migrates inwards as the protostar becomes less luminous and contracts. Closer in towards the protostar temperatures are far higher, and these volatile gases are more widely dispersed. Far beyond the snow line the density of gases declines rapidly, and it becomes more difficult to form very large, gaseous planets.

Thus the expectation was that Jupiter-mass worlds would form close to the snow line of their parent star. Yet, if this was true, how did Peg 51b end up so close to its star? It could have been that 51 Peg b was some unique object, with no counterparts in other star systems. However, in the years that followed its discovery a succession of other hot Jupiters were found, confounding the problem. Since these planets could not have formed in situ they must have migrated inwards. This raised immediate questions. How was such extensive migration possible? Did it occur in every planetary system, including our own? If planetary migration is common, why doesn't our planetary system have a hot Jupiter?

In 1979, thoroughly presaging the discovery of hot Jupiters, Peter Goldreich (Caltech) and separately Douglas Lin (University of California, Santa Cruz) proposed that planets could and would migrate as they were forming around their star. Two types of migration were considered: Type I and Type II. Type I migration was caused by the planet orbiting within any residual gas and dust in the system leftover after planet formation. As the planet orbited within this detritus frictional drag would cause it to spiral inwards and scatter the gas and dust outwards. Type II migration would occur once the planet had grown in mass and cleared a path through the debris disc. Lin suggested that material would still stream from the edges of the disc onto the nascent giant. Due to the distribution of the material in the disc, the planet would continue to migrate inwards while a wave of denser material within the disc transported angular momentum outwards. This latter mechanism, rather than the first, has been invoked to explain the presence of most hot Jupiters.

However, this is not the whole story. Analysis of the orbits of hot Jupiters reveals that many (perhaps 25–50 %) have orbits inclined to the equatorial plane of the star system. Konstantin Batygin (Harvard-Smithsonian Center for Astrophysics) analyzed the orbits of these unusual worlds and concluded that the most likely scenario for the formation of these unusual orbits was the early influence of additional suns. Although many of these planets are hosted by a single sun at present, most models for the formation of star systems posit that they are born multiple. In such a situation the varying and often chaotic gravitational influence of these additional stars would alter the orbits of the planets that are forming. In some instances Jupiter-mass worlds would be scattered inwards, through the protoplanetary discs, leaving them in hot, tight orbits close to their parent Sun. Systems such as ours and Gliese 581, with a single sun, may have begin life with more than one, but later this additional star (or stars) was removed through gravitational encounters with other bodies.

The fossil record in each case is the inclined and sometimes retrograde orbit of the planets that are left. Indeed, the orbits of the planets in what appears to be an orderly solar system bears witness to earlier carnage. In our system planets have a spread of orbital inclinations, up to 7 % from the equatorial plane of the sun, suggesting that we, too, may have had a second or third star lying more distantly that was then "lost."

For whatever reason each surviving hot Jupiter paused short of falling into its star and ended up locked in a screechingly fast and tight orbit around its star. Indeed so abundant were these hot Jupiters that their presence demanded such a scheme. Nearly 20 years after the discovery of 51 Peg b, most astrophysicists now readily accept the Type I or II model to explain the presence of hot Jupiter worlds.

The Nice Group

So what of our Jupiter? Was it always in such a sedate, distant orbit around Sol? Apparently not. Close examination of the distribution of asteroids within the main belt paints a portrait of family violence in which sizable and often abrupt phases of planetary migration occurred – some of them seemingly fairly protracted in nature. Moreover, given the predicted availability of material around the Sun Mars should have been bigger than it is. Did something prevent Mars from growing to greater dimensions?

In two broad washes of work in 2007 and 2011, the group of Alessandro Morbidelli, Harold Levison and latterly Kevin Walsh (part of the "Nice Group" as they are collectively known) published a slew of articles that neatly explained the location of different types of asteroids in our Solar System and the low mass of debris that constitutes the Kuiper Belt. Although not proven, the work does tie up some peculiar loose ends – the location of the Trojan asteroids that accompany Jupiter; the mixed distribution of S-type (basaltic) and C-Type (chondritic) asteroids; as well as the diminutive stature of Mars.

In 2007 Morbidelli's group published a string of articles that examined the period in the history of the Solar System called the late heavy bombardment. The analysis suggested planetary migration was the source of this cataclysm. A few years later Kevin Walsh (University of Nice, Sophia-Antopolis) published a groundbreaking article in *Nature* that suggested that Jupiter and Saturn had undergone an unexpected two-way migration much earlier in their history. This earlier phase not only solved the Martian mystery but also resolved oddities in the distribution of some of the asteroids in the main belt.

The first phase occurred within 100,000 years of the formation of the Solar System. At this stage the proto-Jupiter had amassed 60 times the bulk of Earth and lay 3.5 A.U. from the Sun in what was then the snow line. Soon thereafter, it began to migrate inwards, drawn by the mass of gas and dust that lay close to the developing Sun. It was here that it would affect the fate of the inner planets, Mars, Venus, Mercury and Earth. As such, Jupiter was pursuing the same pattern of inward migration that is proposed to account for 51 Peg b. At this point the terrestrial planets were no more than a twinkling in the Sun's eye, granules of olivine and other silicates and gas, but the effect of this migration would be profound (Fig. 9.3).

Having already accumulated a sizable proportion of its mass, Jupiter exerted a considerable force on its surroundings. What was not absorbed into its mass was spewed out of the disc in every direction. This enriched the inner Asteroid Belt with silicate-rich and relatively dry planetesimals and also scattered a small proportion of them out into the outer Asteroid Belt.

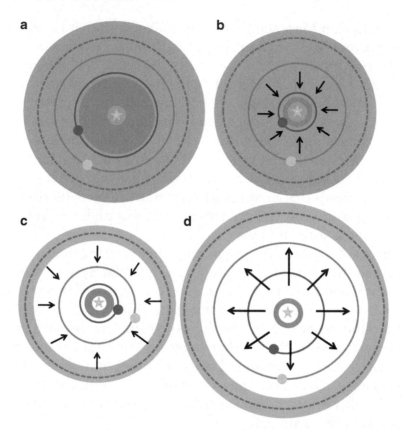

FIG. 9.3 The orbital migration of Jupiter and Saturn. Between 150 and 500,000 years after initial formation, Jupiter (*red dot*) moves inwards from 3.5 to 1.5 A.U. Conserving angular momentum, material inwards of this is scattered outwards (**a** to **b**). At (**b**) Saturn (*orange dot*) also migrates inwards, scattering material as it goes. Eventually it catches up with Jupiter (**c**). Interactions between both cause them to move back out to near where they lie now (**d**). The *blue* areas represent zones populated by planetesimals. The darker the blue shading, the more massive the debris present

Somewhat further out, near 4.5 A.U. from the Sun, Saturn was also coalescing from the disc of material, and by 50,000 years of age had accumulated 30 times the mass of Earth. As Saturn assembled Jupiter had migrated inwards to where Mars currently resides at 1.5 A.U.

Within the remaining disc of debris a property called resonance was quickly established. Resonance occurs when one massive object completes a single orbit for every two or more that a more distant body does. In essence as the massive body orbits its star, it will pull or push on additional objects located at specific distances from it, through the force of gravity. These forces gradually accelerate the second body in its orbit so that it moves outwards. These resonances are particularly effective when there is a ratio of 2:1 between the interacting bodies: the inner body completing two orbits for every one of the more distant world. Such a resonance was then set up between the orbits of the remaining material located near the Sun and the inward migrating Jupiter. These drove a large proportion of this matter into a narrow disc located between 0.7 and 1 A.U. from the Sun. It was here, much later, that Mercury, Venus and Earth would eventually form. Mars was still a dream away.

Why did Jupiter stop at the current orbit of Mars? It appears that the developing Saturn was key to this. Although getting off to a more hesitant start, Saturn eventually accumulated enough mass to begin its phase of interaction with the same debris disc that Jupiter had done millennia before. After 100,000 years it, too, began to move inwards, eventually encroaching on the destructive Jupiter. Jupiter and Saturn then developed a 3:2 resonance – Saturn completing two orbits for every three of Jupiter. This accelerated the more massive Jupiter under Saturn's influence. Jupiter then promptly reversed its approach on the Sun and began to move back outwards while Saturn continued to pull away from Jupiter like the teasing rabbit on a greyhound track.

Once again, debris lying outside the orbits of these planets were thrown both outward and inward. This time it was the carbonaceous C-Type asteroids that were flung inward. By the time Earth and other inner rocky planets were forming, the material in their vicinity was highly enriched with icy, water-rich material. Thus Earth, although relatively dry by the standards of Uranus and Neptune, still has abundant water. By 600,000 years after formation both Jupiter and Saturn had reached positions close to where they are now, and things temporarily settled down.

Lying in more distant orbits at 6 and 8 A.U., Uranus and Neptune were shunted further outwards from the Sun, reaching positions at 10 and 13 A.U., respectively. Saturn at this point lay at 7 A.U. It still had some way to go, but its final maneuverings would prove important for the later development of life on Earth. Although models don't clearly resolve this issue of the two outer planets, it is possible at this stage that Neptune's orbit lay inside that of Uranus'.

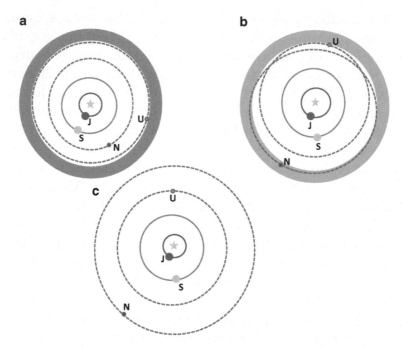

FIG. 9.4 The Late Heavy Bombardment. After formation (**a**) the giant planets lie inwards of a massive disc of debris. At (**b**) the slow outwards orbital migration of Saturn (*yellow*) kicks Uranus (*blue*) and Neptune (*green*) out into the disc. Most of the disc is scattered outwards, but much falls in towards the Sun, blasting the inner planets. At (**c**) (3.9 GYa) the current arrangement of planets is achieved. In some but not all models, Neptune begins its life inside the orbit of Uranus (**a**)

What of Mars? Finally, after this first phase of migration was over, Mars could form from a depleted store of material not already swept up by Earth, Venus and the outer gas giants. Hence Mars was left small. There simply wasn't enough material left to form a larger world. A shame for us, as a larger Mars would undoubtedly be habitable today. Then again, a larger Mars might just nudge Earth into a tighter, and hotter, orbit around the Sun (Fig. 9.4).

The Late Heavy Bombardment

By 50 million years the Solar System's worlds were all formed, but things were not yet settled. Once again, it was when the researchers examined two populations of asteroids that they realized that the outer planets hadn't settled into their current orbits until much later in the Solar System's history than was first thought. Alessandro Morbidelli of the Nice Group (Observatoire de la Côte d' Azur) proposed a second wave of planetary migration 500 million years after the Sun had formed. One of these groups of flotsam and jetsam was the Trojans, lying along Jupiter's orbit. The other asteroids resided in the main belt, between the orbits of Mars and Jupiter. Once again this jostling was driven by the migration of Jupiter and Saturn.

After Jupiter reached a location close to where it is now, it entered a new 2:1 resonance with Saturn, completing two orbits for every one of Saturn's. The initial drift had been slow and took nearly 500 million years, but subsequent realignments would be much faster. As Jupiter and Saturn interacted once more, the more distant Saturn was driven swiftly outward, migrating to its current location at 11 A.U. In order to conserve angular momentum, Jupiter was compelled to move slightly inward to its current location at 5.2 A.U.

As Saturn pulled away from the Sun it then entered a phase of interaction with the ice giants Neptune and Uranus, which lay next out from the Sun. This phase was also mercifully brief. Saturn kicked both Neptune and Uranus outward, plowing them through the disc of icy planetesimals that lay in the outer fringes of the Solar System. It is conceivable, at this stage, that Neptune lay between the orbits of Uranus and Saturn, but once the process began Neptune was booted past the orbit of Uranus and out toward the suburbs of the Solar System. Not all of the model simulations run by the Nice Group show this, but it is possible.

At around 4,000 million years ago, or 550 million years after the formation of the Solar System, the planetary carnage began. As Neptune plunged through the outer disc of planetesimals, thousands were scattered inward and outward. Once more, a rain of icy material swept through the Solar System towards the Sun. Earth and Moon, and presumably the other terrestrial planets were pummeled. Although few Earthly signs remain, the Moon still wears its scars forlornly. The Late Heavy Bombardment, as this period in geological history is known, was brief, perhaps 200 million years in duration. However, its effects were undoubtedly profound. Given the larger surface area of our world, and its greater gravitational pull, Earth must have been something of a wreck by the time the majority of the debris was cleared away.

The apparent brevity of the Late Heavy Bombardment (LHB) may be misleading. Recent analysis by Morbidelli (Nice Group), Simon Marchi, William Bottke (both of the Southwest Research Institute, Colorado) and David King (Lunar and Planetary Institute, Houston) indicates that although the bulk of the collisions occurred in a narrow window between 3,900 and 3,800 million years ago, a significant number of massive impacts occurred as recently as 1,800 million years ago. Many of the impactors were tens of kilometers across and could well have come close to sterilizing the surface of our world.

Life clearly survived, but close scrutiny of the evolutionary record may identify the scars from these collisions. It is even conceivable that during the Hadean era, from 4.556 billion to 3,900 million years ago, life arose repeatedly only to be eliminated in vast, sterilizing impacts, in what were part of a more continuous rain of fire than the record of the LHB implies. Given the apparent rapidity at which life's arose following the end of the phase of most intense bombardment, it suggests that life either arose much earlier, somehow surviving each apocalypse, or more perhaps likely, arose quickly at the end of the Late Heavy Bombardment. Moreover, if the rapid emergence of life is real, it suggests that the development of life is a very probable event and should arise quickly on whatever worlds have both an energy source and liquid water. Were life difficult to create, it would not have arisen so promptly after the cessation of the Late Heavy Bombardment.

How does this simulation explain the presence of the Trojans? Moreover, how did they obtain their odd properties? The Trojans are peculiar objects. Given their distance from the Sun they should be very icy, with a composition dominated by water. Instead they appear to be relatively dry objects, locked in orbital resonances at 60° angles along the path of Jupiter. The Nice Group model suggested that these objects began their lives further out in the cold Kuiper disc. As the outer planets migrated, many of these icy bodies were thrown inwards. Those that escaped pummeling the inner planets, or diving into the Sun, were thoroughly baked, driving off much of their volatile content. What was left was more rocky, carbonaceous bodies. Many of these depleted bodies were eventually spun back outward to be captured by Jupiter's immense gravity. Residing at loci focused on 60° angles to the massive planet, the gravitational pull of Jupiter is balanced by that of the Sun. The Trojans had finally found a stable home far from the Sun.

The processes of planet formation and migration are unlikely to be unique. Models of planetary formation going back four decades suggest a very messy process, with multiple rounds of collisions, planetary migration and orbital swapping. Evidence from extrasolar planetary systems suggests that these processes are universal, or at least very common. Given the wealth of material left over after planet formation, it is likely that all inner terrestrial worlds will be endowed with volatile materials such as water soon after birth, even if they were relatively dry at inception. For example our Earth must have been baked thoroughly dry early on. Even if nature had bequeathed our world with a rich tapestry of volatile materials at birth, the catastrophic formation of the Moon, through a giant impact, would have driven most of this back out into the vacuum of space. The present bounty of water – or most of it – must have been delivered after the formation of our Moon.

The Kepler telescope has provided further outlandish evidence that supports the idea that planetary migration is not confined to our homeland. The Kepler-36 system is a case in point. Kepler-36 has two closely paired worlds in neighboring orbits – nothing unusual there, you might think. However, each planet orbits its Sun at approximately the same distance as Mercury does ours. Note that this is *each* planet. Kepler-36b orbits its star at 0.1153 A.U. in 12 days, while Kepler-36c at 0.1283 A.U. with a period of 16 days. This means that the distance between them is less than 10 % of their Sun-planet distance. This is remarkably close.

More remarkable still are their compositions. Kepler-36b is a super-terran planet with 4.5 times the mass of Earth and likely rocky, perhaps with a deep ocean of water. Kepler-36c weighs in at eight times Earth's mass and is probably a scaled-down version of Neptune.

Despite their extreme proximity, the transits of these planets reveal that there is an eightfold difference in density, clearly indicating a very different formation history. The only way each of these "chalk-and-cheese" worlds could have ended up in this configuration is if the outer mini-Neptune-like world migrated a long way in toward its new partner. There simply could not have been such a gross difference in the density in such a narrow region of space in which each is now located. They must have formed in different neighborhoods.

The central star is yellow G-class star with marginally more mass that the Sun. Their star is also slightly older than the Sun, meaning it is also hotter for its mass than a star of the Sun's age. Given the proximity of both planets to their star, neither is likely a host to life.

Putting the processes of orbital migration in context, even if the formation of Gliese 581 initially led to dry super-terrans, orbital migration would, most likely, drive icy material towards the parent star, endowing the planets with their own supply of water and other volatile materials. Indeed there are indications from the present orbits of Gliese 581's worlds that Planets b and c migrated inwards. Such resulted from interactions between the planets themselves, or each of the tightly orbiting worlds formed further out, but interactions between each world and the residual disc of material dragged them inwards. If this is true, having formed at a greater distance from their host star would ensure that each was born with an abundance of volatile materials, such as water. Indeed, Planet b seems too large to be entirely rocky. At 15–30 Earth-masses, much of this mass must be gas or water vapor.

A low metallicity star such as Gliese 581 is even less likely to contain the necessary abundance of heavy elements needed to form entirely rocky, Earth-like worlds of the sizes seen around Gliese 581. Much of Gliese 581b's bulk must be composed of more volatile materials. In turn, this implies that this world must have formed further out, beyond the snow line of the infant star, before migrating inwards.

Putting Gliese 581d's Formation and Early History in Context

Turning our attention to the potentially habitable Planet d, its size of 5–7 Earth masses also implies a fairly substantial amount of volatile material. These compounds will strongly affect the habitability of the world, both in terms of providing the raw materials for life, but also on the geological processes that are necessary for the sustenance of life.

Imagine a world newly formed. Ninety percent of its mass is rock and metal. The remaining 10 % is volatile material such as hydrogen, oxygen (not bound as metal oxides), carbon, nitrogen, sulfur and other light gaseous materials. Most of these elements, which are not already combined with metals, end up combining with themselves to form gases such as water vapor, carbon dioxide, sulfur dioxide, hydrogen sulfide and ammonia. Of these, the fate of ammonia is important. Deep within the hot mantle of the new world, ammonia (in a charged form called an ammonium ion) has been mistaken for the similarly sized potassium ion. This case of mistaken identity traps it within silicate compounds. Under the extreme conditions of the planetary interior, ammonia is compressed and baked until it breaks down into its component hydrogen and nitrogen atoms. The nitrogen atoms combine in pairs and escape through volcanic vents into the depths of the deep ocean above, while the hydrogen reacts with metal oxides to form water. This, too, is light, and the heat drives it outwards through volcanic vents. Soon, any mantle water is bubbling outwards and combining with icy material on the surface of the planet. A deep, salty ocean, bathed in a nitrogen-rich atmosphere, is born.

If the amount of available hydrogen (or ammonia) is great, nitrogen will remain dominantly in the form of ammonia. The abundance of ammonia is an important consideration for the formation of life and for the formation of an oxygen-rich atmosphere. Although ammonia is a useful building block for biomolecules, its presence in large quantities produces very alkaline environments. These could preclude the development of life by reacting with, or destroying, important molecules. Long-chain biomolecules such as proteins and nucleic acids are unstable at a high pH, and although organisms on Earth can adapt to these conditions, high alkalinity might well prevent the initial development of living organisms before life has a chance to adapt. Moreover, abundant ammonia would readily react with oxygen and inhibit its build up in the atmosphere, assuming oxygenic photosynthesis (Chap. 8) can even arise.

There is undoubtedly a critical cut-off in planetary mass where ammonia dominates over free nitrogen. Once the abundance of silicates becomes less than the abundance of ammonia there will not be enough to soak up, constrain and ultimately transform in the planetary depths. The transition from terrestrial world with its nitrogen atmosphere to more icy worlds dominated by ammonia, methane, hydrogen and helium presumably lies near the snow line, where the abundance of these gases increases sharply. Planets formed outside this line are therefore unlikely to be terrestrial in nature or have atmospheres dominated by nitrogen.

In the Gliese 581 system planetary formation was largely complete within 50 million years, this despite the ongoing sluggish process of star formation. The final stages of planetary assembly involved the coalescence of large planetesimals. This is a critical and highly unpredictable process and can lead to very different outcomes. Where the planetesimals that formed Gliese 581 were largely differentiated – that is, had separated into distinct layers according to density (Chap. 2) – violent collisions between planetesimals might vaporize much of the accreted volatile material into space, leaving relatively dry worlds. This scenario is explored in the next chapter, where we look at Gliese 667Cc. In this section we explore the development of a planet endowed fully with a modestly thick blanket of water and gas.

Given our relatively mediocre understanding of this planetary system, the concluding sections of this chapter are inherently subjective and mostly guesswork. This is all the more true given nature's propensity of demolishing the best of our naive suppositions with cruel, hard facts. In order to flesh out probable conditions on Gliese 581d we will scale Earth up and extend the mass range to bodies within our Solar System. This is reasonable in the sense that the underlying physics is probably universal, based on current knowledge. However, it may be utterly naive given the size of our known working sample: one. The basic idea is that planets will be dominated by iron and silicates up to 3–5 Earth masses. Beyond this the material available to produce worlds will be increasingly dominated by more volatile, and less refractive materials – so-called ices such as ammonia, water, and carbon dioxide.

We assume that the red dwarf system is merely a scaled down version of our own, and this, as we have seen, has its limitations. If so, the bulk of material in the disc from which the planets formed is hydrogen and helium. In Gliese 581 at most 1 % of the stellar material is comprised of elements heavier than hydrogen and helium. The abundance of these elements scales with the process of nucleosynthesis in stars. We already know that the majority of these heavy elements are carbon and oxygen, with nitrogen, iron, silicon, and elements such as magnesium and calcium pulling up the rear. Based on the trend for more iron-rich debris with time and some knowledge about the parent star, we have a good starting point for the planets that surround it.

Given the relatively low mass of the protostellar nebula that gave rise to the Gliese 581 system, a smaller fraction of material was available to form planets. Jupiter-mass planets were predicted to be rare in such circumstances by astrophysicists such as Greg Laughlin. Observations by researchers have largely confirmed this sparsity, with at most 10 % of all red dwarf worlds being Jupiter-like. Instead smaller worlds dominate the red dwarf camp. However, despite the difference in sizes, the mechanism giving rise to them likely scales in a similar if not identical manner to our Solar System.

Given all these constraints worlds with masses exceeding a few times that of Earth are likely to contain proportionately more light elements. Thus at some critical juncture between 1 and 7 Earth masses volatile materials, such as water, are likely to become dominant. If elemental proportions remain similar in all protostellar nebulae, then there is probably no way around most super-terrans having surfaces dominated by water. Lowering the proportion of metals probably makes the situation worse, by decreasing the availability of elements heavier than oxygen. Thus we should remember that both Gliese 667 and Gliese 581 are relatively metal-poor stars compared to the Sun. This, in turn, implies planets more likely to be dominated by lighter elements.

If we assume a "best-case scenario" one Earth mass of a seven Earth mass planet might be water and other light compounds or elements. Depending on the nature of the other heavy elements, this would lead to a planet with a diameter two to three times that of our world and a surface gravity of between 1.5 and 2.5 that was experienced on Earth's surface. (The precise value depends on the diameter of the planet.) The greater the mass of water and light elements that are present, the lower the planet's density and the greater its diameter. If we build a planet with six Earth masses of rock and metal this will be approximately twice the diameter of Earth. Add an additional Earth-mass of water and light gases, and the diameter will increase substantially more than if this was additional rock. The lower density veneer would add a substantial depth of ocean and atmosphere on top.

Transits are the best source of planetary data, as they can give planetary mass and radii – as well as allowing an investigation of any atmosphere. In future high-resolution infrared data might substitute for the latter point as it has done for τ-Bootes b, but for now transits provide the greatest wealth of information.

If we go with a 6:1 proportion of heavy to volatile elements then we would get a world with a massive rock-metal core bathed in an ocean hundreds of kilometers in depth, capped with a deep atmosphere dominated by water vapor and nitrogen. Although perfectly habitable in terms of temperature and composition, the deep atmosphere would most likely produce such formidable pressures to preclude all but the simplest life forms, if any at all.

On Earth, there is good evidence that life started at or near hot springs and then migrated to balmier abodes. However, if the ocean is tens to hundreds of kilometers in depth, pressures might be so severe as to preclude the spread of even simple biology from the ocean depths. Consequently, unless circulation was very efficient and the atmosphere peculiarly transparent to visible light and, importantly, infrared radiation, photosynthesis would likely be impossible. Such a deep ocean world would not be a pleasant place to visit. Life, did any arise, would likely not progress beyond microscopic forms.

Yet, for anything able to survive these conditions, the environment would be unlikely to change over most of the life of the star system. The only measurable change would be the remorseless decline in geothermal activity as the planet cooled down. In the end, it would be the faltering internal clock that killed off life, rather than anything occurring in the atmosphere or the star above (Fig. 9.5).

Another important factor is the effect of such a deep ocean on the process of plate tectonics. A very deep ocean will also produce very high pressures at the depth of the solid crust. Assuming that plate tectonics is driven by the formation of eclogite (Chap. 4), then the transition from basalt to eclogite would occur at very

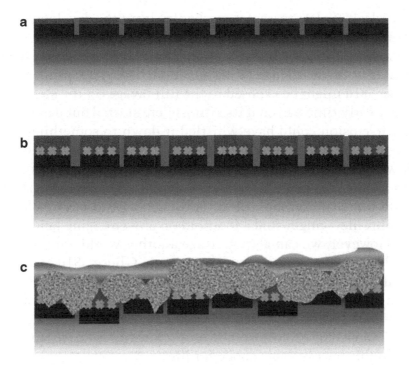

FIG. 9.5 The rapid formation of a thick lid of continental crust on a young, hot super-terran. Extensive production of basalts (**a**) produces a very thick crust. The thick crust and higher surface gravity causes the base of the crust to founder into the underlying mantle, where it melts (**b**). The resulting sodium-rich granites form a lid that can resist subduction (**c**). Plate tectonics ends early, and the cycling of carbon dioxide to and from the mantle ends. This dooms most photosynthesis to a premature end and limits the habitability of the planet

shallow depths in the crust because of the weight of water above. This would probably cause the crust to founder very quickly. The great pressures might even prevent the generation of basalts altogether. In turn, this would prevent efficient plate cycling and could lead to either the formation of a thick lid of rock that resists cycling into the interior altogether, or results in the formation of a thick lid of granite that has the same effect at very early times. Either way, without efficient cycling of crust, the availability of carbon dioxide and other raw materials would be restricted. The ocean would end up largely barren. Life wouldn't get started or progress far. Therefore, the depth of any global ocean is likely to be a significant factor in the subsequent development of life.

A lack of plate tectonics might also limit the development and maintenance of a global magnetic field, particularly as the planet will rotate slowly. This would have longer term impact on the sustainability of the planet as an abode for life. The atmosphere of Gliese 581d might be exposed to the full ravages of the stellar wind at very early times. Even if its atmosphere started out dense, atmospheric erosion could have whittled it down to something resembling that of Mars in the 7 billion years the planet has existed. This allows for an interesting caveat to the rather negative outcome discussed above. If the atmosphere is eroded, the depth of the ocean would decrease. Perhaps, then, as the global ocean shrivels, life – complex life – might find a foothold on any emerging land.

However, we can also envisage another world with (possibly) utterly unrealistic qualities. Let's assume Gliese 581d is an ocean world, but not one with a deep, Earth-mass of water pounding down on the rock below. Instead, Gliese 581d is predominantly rocky but has a global ocean measuring no more than 10 km in depth, capped by an atmosphere only a few times denser than ours. Would this be a suitable abode for life?

The short answer – yes.

If we insert a rocky world with a suitably thin overlying ocean, then the biological processes we are used to can operate. Magmatism and plate tectonics cycle carbon dioxide from the ocean to the mantle and back again. The composition of the atmosphere is then dependent on that of the ocean, rather than an abundance of hydrogen accreted from interstellar space. The atmosphere will initially be dominated by carbon dioxide, water vapor and nitrogen gas. Over time, without access to the volcanic processes occurring on the floor of the ocean, the atmosphere will steadily become depleted in carbon dioxide.

As we saw in Chap. 5, carbon dioxide dissolves in water, forming carbonic acid. As this rains into the global ocean much of it will react with dissolved calcium and magnesium ions, or be taken up by living organisms. In consequence carbon dioxide salts will progressively rain out onto the ocean floor far below. Carbon dioxide will only be replaced by slower processes of respiration (if it occurs) and the steady escape of carbon dioxide from the ocean. In turn this is dependent on the ocean temperature and how much is supplied by volcanic processes at its base. If temperatures are low, carbon dioxide will tend to stay in solution, and the rate of respiration of any living organisms floating in the ocean will also be low. Thus over time, the amount of atmospheric carbon dioxide will fall, probably at a faster rate than here on Earth, where surface volcanic activity replenishes it.

Assume the oceans are not too deep and that life can start before the central star has reached the main sequence. Gliese 581d is thus afforded an opportunity to spawn living organisms. Given Gliese 581d's marginal position in the star's habitable zone, it is conceivable that Gliese 581 would then have begun warm and habitable, with life developing in the ocean depths. As Gliese 581 approached the main sequence, the stellar flux of radiation declined. However, assuming conditions were clement initially, the declining load of radiation meant that the stellar habitable zone contracted inwards, and the amount of radiation received by Planet d fell. As temperatures subsided, the solubility of carbon dioxide gas in its oceans increased. After a few hundred million to a few billion years, the loss of atmospheric carbon dioxide led to the planet-wide ocean freezing over.

In this frozen state, with only the slowest evolutionary push from its small, central star, water-world Gliese 581d may well have frozen over billions of years ago. In this state life would have become restricted to a thin biosphere trapped in a shallow veneer of liquid water at the base of its global ocean. This could be sustained only by the declining tick of the geothermal clock. If Gliese 581d had frozen, it might resemble a giant version of Europa. This ice world would be uninhabitable on its frigid surface, but might well remain alive within its hidden ocean depths. Life would continue for billions of years until the loss of internal heat shut down the supply of nutrient-rich waters. With no means of escape through the dense ice shell, life would then crumble and slowly go extinct.

Perhaps this is unnecessarily gloomy. With the radiation from Gliese 581 dominated by infrared, the buildup of snow and ice has a far less pronounced effect on global temperatures than visible light has on our world. On Earth the Sun's light is readily reflected by snow and ice, and changes to the amount of land covered by these materials can cause a significant drop in global temperatures. During the period from 750 to 650 million years ago

Earth appears to have largely frozen over several times. Only the steady exsanguination of carbon dioxide from volcanoes appears to have arrested the process.

For planets orbiting red dwarfs the story is a little different. Infrared radiation is less readily reflected by snow and ice than visible light, and so such an icy runaway can be avoided. Gliese 581d may thus be a tad slushy rather than completely frozen. A slush ball is a perfectly habitable place for a variety of organisms, including those that are complex and require oxygen. The limiting factor will be the productivity – the amount of usable food – the oceans can produce. If this is still high, complex webs of life will extend throughout its depths and the planet will be vibrant. However, if the atmosphere and oceans have run critically low in carbon dioxide, the food chains that support any higher life-forms will have collapsed and life may either be extinct or teetering on the brink of collapse.

Either way, if Gliese 581 is 7 billion years old, life may at best simply be clinging on. It would seem unlikely that this planet will currently sustain anything more complex than organisms we might call bacteria.

The Fate of the Gliese 581 System

As Chap. 3 explored, red dwarf stars don't rush at anything. Their lives are a sluggish crawl from cradle to grave. The universe is 14 billion years old, but not one M-class star has even made it out of kindergarten. The most massive M0 red dwarf with a mass of 0.6 solar masses will expire in another 80–90 billion years. The least massive 0.075 solar mass M8.5 or L1-class dwarf will last over 10 trillion years. There is no hurry. However, death gets us all in the end.

In the first part of this section, the fate of Gliese 581 and its planets will be discussed. It will be a slow process with seemingly nothing happening for much of the history. When events finally take a turn for the worse, Planet d will be over 350 billion years old and undoubtedly showing its age.

The age of the star system is highly uncertain. Ages range from 2 to 11 billion years. The lower estimates are unlikely given the

Table 9.1 The properties of Gliese 581 as it ages

Age (billion of years)	Surface temperature (°C)	Diameter relative of sun	Luminosity relative of sun
10	3,310	0.303 (30.3 %)	0.0122 (1.22 %)
20	3,310	0.304 (30.4 %)	0.0135 (1.35 %)
50	3,310	0.309 (30.9 %)	0.138 (1.38 %)
100	3,290	0.318 (31.8 %)	0.0144 (1.44 %)
150	3,280	0.322 (32.2 %)	0.0151 (1.51 %)
200	3,250	0.342 (34.2 %)	0.0158 (1.58 %)
250	3,200	0.360 (36.0 %)	0.0166 (1.6 %)
300	3,150	0.385 (38.5 %)	0.018 (1.8 %)
350	3,130	0.422 (42.2 %)	0.0212 (2.12 %)
400	3,180	0.515 (51.5 %)	0.0378 (3.78 %)
450	3,730	9.04 (904 %)	19.2 (1,920 %)

low metal content of the star. In general metal content decreases with age. Population II stars born, soon after the universe's birth, have very low metal content – perhaps as low as one millionth that of the current Sun. Stars born today have metal content several times that of the Sun. The best estimates place Gliese 581 at between 5 and 7 billion years old – somewhat older than the Sun. The age is important, as several geological systems may be showing their wear and tear, even though the parent star is in its comparative infancy.

Table 9.1 illustrates the slow pace of change of stellar properties. These are adapted from work done by the group of Kevin Lee at NAAP (The Nebraska Astronomy Applet Project at the University of Nebraska-Lincoln).

Over tens of billions of years helium slowly accumulates within the star. There is likely to be a small radiative core surrounded by an envelope of hydrogen and helium that is constantly stirred by convection (Chap. 3). Were Gliese 581 slightly less massive the star would most likely convect throughout its bulk, ensuring a steady supply of fuel to the stellar core. However, with a mass of 0.3 solar masses, Gliese 581 may not benefit from this trick.

Overall, for the first 350 billion years, the star slowly grows in diameter and brightness, and this is accompanied by changes in its surface temperature. This mirrors changes in more massive stars like the Sun.

After 350 billion years Gliese 581 begins a very protracted phase of expansion to become a red giant. At around 400 billion years, Planet d is finally stranded on the hot side of the habitable zone and any oceans evaporate, probably within a few million years. The time depends sensitively on the depth of water and the thickness of the atmosphere. The expanding surface of Gliese 581 begins to encroach on Planet d, and the atmosphere will begin to be blown off into space. Earth, by comparison, is stranded on the hot, Venusian side of the Sun's habitable zone a good 4–5 billion years before the Sun expands into a red giant.

Proportionately, Earth will have spent a greater period of its star's main sequence time uninhabitable than Gliese 581d. The total habitable time for Earth will be approximately 5.5 billion years out of roughly a 10–11 billion year main sequence life. By contrast Gliese 581d will be habitable for 400 billion out of its 450 billion years on the main sequence. From where we are now a 10 % increase in luminosity will be all that is needed to extinguish life on Earth. By contrast a 10 % increase in the output of Gliese 581 will make the most trivial of changes to Gliese 581d. Earth has already, and will continue to, see the Sun increase in luminosity by 40 % during its period of habitability. Although Gliese 581 doubles in brightness during its 350 billion years of stability, this is from such a low baseline that the stellar habitable zone barely moves until the star is close to death. Thus planets within the habitable zones of red dwarfs experience the most prolonged phase of habitability – at least in terms of their parent star's contribution.

After nearly 450 billion years – or more than 30 times the current age of the universe – Gliese 581 reaches the tip of the red giant branch. Fortunately, for Gliese 581d, a red giant made by such a small star is considerably less expansive than the one our Sun will become. Whereas the Sun's radius will extend to reach 150 million kilometers, the expanding Gliese 581 will peak at less than ten times that of our present Sun – barely 10 million kilometers. Were this not the case the entire planetary system would be destroyed. Instead, Planet d is thoroughly baked, but escapes vaporization within the swirling envelope of the star. However, innermost Planet e is boiled away, as is Planet b. Planet c will escape, coming perilously close to annihilation with an orbit only 1 million kilometers or so from its star's bloated surface.

With a mass well below the limit needed to ignite helium, the core ceases contraction when its density reaches one million times that of water at Earth's surface, and its temperature has climbed to less than 30 million degrees. For a few million years, the star will continue to fuse hydrogen into helium through the carbon-nitrogen cycle in a shell surrounding its core (Chap. 3). This will keep it shining with a luminosity close to that of the present Sun.

For Gliese 581 the end is in its beginning. As Gliese 581 expands, stellar winds will begin to strip hydrogen and helium from the outer layers. Indirectly, this can be seen in the data presented in the table. Initially, as the star expands, its outer layers cool down and the surface temperature declines. However, as the outer layers are thinned away by stellar winds, progressively hotter layers are exposed. You can see the beginning of this phase at 450 billion years as the surface temperature begins to rise. Shortly after peaking, the star begins its brisk final stage. Whereas earlier evolutionary stages were measured in tens of billions of years, the final installment in the star's life lasts less than one million. With a dwindling supply of hydrogen fuel and a core unable to evolve further, the star peels away from the red giant branch, and quickly slides leftwards across the Hertzsprung-Russell diagram and onto the white dwarf cooling tract. As the core contracts and reaches planetary dimensions, the stellar surface heats to 80,000°. With the temperature climbing, the bulk of the radiation that is emitted shifts from the visible to the ultraviolet. This energetic radiation fluoresces the expanding cloud of hydrogen and helium, encasing whatever planets remain in an expanding planetary nebula.

Not long after Planets b and e are destroyed, whatever remains of the system begins a final protracted phase of cooling. The core of Gliese 581 becomes a helium-rich white dwarf of 0.28 solar masses. All traces of former planetary habitability vanish. For nearly the rest of eternity, Gliese 581 cools into invisibility, and another chapter in the habitability of our galaxy closes.

Conclusions

The final call on such super-terran water-worlds depends to an extent on how long plate tectonics can continue, and this depends on mass. Beyond three Earth masses, calculations suggest that the greater the mass the quicker the surface becomes swamped with granite and the quicker plate tectonics is choked off. If Gliese 581 d has a mass in excess of five times that of Earth and it is mostly rocky, we can be fairly certain that plate tectonics will have already ceased, given the age of the system. Life will, at best, be in decline. If we go the other way and reduce the mass of the rock to four Earth masses, and bump up the mass of volatile materials, most likely life would be stifled at birth by the thick atmosphere and deep ocean. Either way it is highly unlikely that a volatile-rich Gliese 581 d is still habitable, except possibly for the most primitive of life forms. Given its age, if it ever was habitable, we may have missed the boat, and instead of observing a vibrant, living world, future missions may simply pick over the bones of a world in decline.

10. The Evolution of an Earth-Like World

Introduction

Lying 22.1 light years from Earth is Gliese 667. Unlike Gliese 581, the Gliese 667 system is a far more interesting construct. For one thing there are three stars, not one. Two of these stars, 667A and 667B, are low mass, K-class orange dwarfs (Chap. 3) orbiting their common center of gravity. The orbit of both K-stars is very eccentric, with their separation varying from 5 to 20 A.U. – roughly the Sun-Jupiter distance at the closest approach, to the Sun-Uranus distance when furthest apart. The stars weigh in at 0.73 and 0.69 solar masses, or roughly three quarters that of our Sun. In terms of size each star more fully approaches that of the Sun rather than Jupiter (Table 10.1).

The evolution of each K-dwarf will be completed in less than 50 billion years, leaving two white dwarfs. This is considerably less time than the red dwarf Gliese 581 will take to expire.

Lying on the fringes of the star system sluggishly orbits a third star, a red dwarf with a mass only slightly greater than Gliese 581. Separated by anything between 56 and 215 A.U. of space, this little star takes hundreds of years to complete one orbit of the central pair. For comparison Pluto orbits the Sun at an average distance of 39.4 A.U., taking 248 years for each orbit. Thus even if we were to swap the far more luminous Sol for Gliese 667A and B, they would still appear as two bright stars in the sky, rather than additional luminous suns as many artistic renderings suggest.

One particularly prominent artistic rendering has Gliese 667C "setting" over the landscape of its habitable and very Utah-like landscape, with the two distant K-dwarfs forming a well separated pair in the evening sky. Quite aside from the lack of sunsets on tidally locked worlds, the separation of the two orange K-class stars

D.S. Stevenson, *Under a Crimson Sun: Prospects for Life in a Red Dwarf System*, Astronomers' Universe, DOI 10.1007/978-1-4614-8133-1_10, © Springer Science+Business Media New York 2013

Table 10.1 The properties of the three stars in the Gliese 667 system

Star	Spectral class	Mass (compared to sun)	Luminosity (compared to sun)	Radius (compared to sun)
GL 667A	K3V	0.73	0.13 (13 %)	0.76 (76 %)
GL 667B	K5V	0.69	0.05 (5 %)	0.70 (70 %)
GL 667C	M1.5 V	0.31	0.0137 (1.37 %)	0.42 (42 %)

is likely too great in these images. When star C is at its maximum distance of 215 A.U. the two distant Suns, separated by less than 20 A.U. would appear as two closely spaced stars. Although probably (just) visible in the daytime sky, they would be nothing more than two bright points of light. Their combined light would approximate a full moon so they could cast shadows but that is all.

Given that stars A and B radiate a combined luminosity of less than 20 % that of the Sun, their great distance from Gliese 667C means that they contribute no meaningful radiation to Gliese 667C's planets and thus have no measurable impact on any living organisms on any of the worlds. As far as any resident of Gliese 667Cc goes, there is only one sun, star C.

Although the artistic renderings of the current system are more than a tad fanciful, as both K-class stars evolve away from the main sequence, the visual impact these distant stars have on Gliese 667Cc will be more profound. Moreover, the death of the two central stars in the Gliese 667 system may also condemn Gliese 667C to a life of solitude, drifting alone in the galaxy (Figs. 10.1 and 10.2).

Measurements of radial velocity by HARPS indicates that the little red star, Gliese 667C, carries with it at least two super-terran worlds,[1] with a further two suspected planets lying at greater distances. An observer on the surface of the outer super-terran would have its red sun permanently fill the sky, with the more distant orange dwarfs relegated to rare, unassuming appearances as a single, bright daytime star.

---

[1] While going to press a team of astronomers led by Guillem Anglada-Escudé of the University of Göttingen, Germany and Mikko Tuomi of the University of Hertfordshire, UK have found evidence for up to seven planets around the star, three of which are potentially habitable. Planet c, described, is the innermost. Planet d, the outermost.

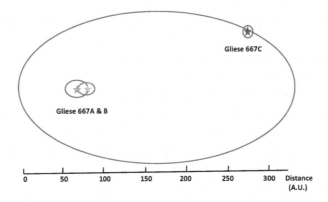

FIG. 10.1 The arrangement of the stellar component of Gliese 667. Stars A and B orbit one another within 20 A.U. However, the planet hosting star Gliese 667C never approaches closer than 65 A.U. and at its most distant is 215 A.U. from the central K-dwarf pair

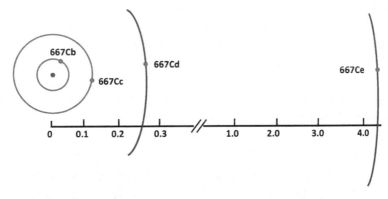

FIG. 10.2 The orbital arrangement of the known and suspected planets orbiting Gliese 667C. Planet GJ 667Cb orbits very closely and is too hot for life, while GJ 67Cc, and possibly GJ 667Cd, are in the stellar habitable zone. A suspected Saturn-mass planet orbits much more distantly at a little less than the Sun-Jupiter distance

The innermost planet, Gliese 667Cb, was discovered by the HARPS team in the autumn of 2009. It has a mass of six Earths, based on their data, but is most likely too hot for life, assuming it even has a solid surface. It orbits its red dwarf sun in a week on an orbit 0.05 A.U. from the stellar surface.

The outer world, Gliese 667Cc, was announced 2 years later and takes approximately 4 weeks to complete one orbit. The world receives approximately 90 % of the energy of our planet, orbiting at 0.12 A.U. from its star. It is, therefore, clearly habitable, at least in terms of the amount of radiation it gets from Gliese 667C. With a mass of 3.9–5 Earths it is most likely a largely solid world, with perhaps a substantial fraction of volatile materials, of which water is probably the most abundant.

If one was to stand on Gliese 667Cc one would see a red sun, somewhat yellower than from the same vantage on Gliese 581d. Although carrying comparable masses, the lower metallicity of Gliese 667C renders the central star somewhat hotter and bluer than Gliese 581. From the planetary surface the star would sit rigidly in one spot in the daytime sky, filling an area more than five times greater than that occupied by the Sun in ours. Standing at the sub-stellar point and staring upwards, the sky immediately surrounding the star might appear relatively blue. However, as you panned your gaze across the sky any blue would rapidly give way to yellower, then red hues. Thus the same issues associated with tidal locking and red dwarfs apply. However, with more abundant and higher energy radiation to go around, Gliese 667Cc is a better candidate for supporting photosynthetic life. With a generous supply of energy from its star even an atmosphere substantially thinner than ours would easily suffice, transferring energy from light to dark sides.

Moreover, the lower planetary mass makes it more likely that the surface of Gliese 667Cc has both solid and liquid regions – continents and oceans. Of course, this is merely more likely, not a proven fact. Although scaling up planets almost certainly means scaling up the proportions of volatile materials, it is not fact, and a number of different variables will come into play.

Although conceivable that a Jupiter-mass planet might be all solid, it would be highly improbable given the proportions of materials available for planetary construction. Overall, more mass probably means more volatiles, and since differentiation separates these from the rock, more volatiles, in turn, means a deeper layer of water and other icy materials on the rocky surface. We might look to build some form of planetary HR-diagram. The lowest mass worlds will be largely solid rocky or icy abodes. With increasing mass, a greater abundance of volatile materials are added, and the planets become first Mars-like, then Earth-like. Add even more volatile material, and they transition through water-worlds to Neptunes and finally gas giants like Jupiter and beyond. Naturally, given the differences in the early evolution of planets, we might have more than one "main sequence" of planets with varying proportions of volatile elements driven by the varying degrees of violence that accompany their formation.

Consequently, if a planet is to be both large and dry, such a world will have to experience some process that depletes it of volatile materials. It could simply be that the super-terrans, which frequent the habitable zones of red dwarfs, are volatile poor, compared to Earth. Certainly, modeling by Jack Lissauer (NASA Ames Research Center) implies this, but volatile-poor is a relative term. A greater initial mass of material might still lead to a water world, even though proportionately less volatile material was initially present than was found on the young Earth. A six Earth-mass super-terran, with only 50 % of the water content per unit mass found on Earth, will still have a deep global ocean and no dry land. Reduce the volatile content further to 15 % that of Earth per unit mass, and some land will be visible, depending on how much water is locked into the planetary crust and mantle. All things are relative, and if we are to have super-terrans with dry land such as Earth, they must be much drier than our world.

There are other options for creating dry super-terrans. We already know the process of planet formation is as catastrophic as it is frantic. Evidence for widespread orbital migration dominates our observations of the Solar System and many exoplanetary systems. In such environments there is plenty of scope for violent collisions between planetesimals. A late, violent collision might then drive off the bulk of a super-terran's volatile materials, leaving a largely dry planet. Although this seems contrived it carries a reasonable probability and may well be an important source of super-terran planets that are habitable for complex land-dwelling life forms. Yet, it may turn out that only planets with masses comparable to Earth (or less) have suitable areas of dry land on which complex, multi-cellular life can develop. Further detailed observations will clearly be needed to confirm or refute this supposition.

Aside from collisions tidal forces could also dry out water-worlds. Io, the innermost Galilean satellite of Jupiter, is effectively dry, despite its location well outside the snow line. All of its water has been driven off by vigorous heating within its interior. Such heating has been forced by the perpetual kneading action of its mass by Jupiter and the outlying satellites. A planet orbiting deep within the gravitational well of its red dwarf star may also experience bouts of violent heating, particularly early on when its orbit may still be adjusting. Rory Barnes (University of Washington), working with James Kastings (NASA Astrobiology Institute) and others, proposed that such heating could be sufficiently strong as to drive much, if not all, of the water content from a developing planet. This will occur if the planet lies in an orbit close enough to its central star. The star itself doesn't directly heat the world through radiation. Gravity does the work, baking the interior until all of its water, carbon dioxide and other biologically useful gases have been driven into space. Although Barnes' work is theoretical, it does, once again, force us to drop an assumption that merely placing a planet in a stellar habitable zone makes it, well, habitable. There are many other factors to consider.

If we are to assume some sort of intelligent life can arise on a super-terran, it would seem likely that it would need dry land on which to do this. Not only does dry land provide a greater diversity of environments on which to develop and acquire sensory information, it also produces a more diverse set of challenges to the underlying genetics (Chap. 7). Evolutionary processes are likely only dynamic where the environment is variable and can provide the necessary selective pressures.

A water-world might allow the development of fish, but fish are not likely to be that intelligent, as the availability of oxygen is restricted by its solubility in water. On Earth at room temperature the air contains 21 % oxygen, but only around 9 % is found in water. Oxygen is likely essential for intelligence. Brains are expensive, fuel-hungry organs to run. Limited oxygen supply will mean that respiration is restricted, body temperatures are likely lower and the processes needed for rapid cell-to-cell communication more muted. Whales and dolphins may be smart, but they are air-breathing mammals and evolved first on land, later returning to the oceans. They do not provide an exception to our aquatic rule. Of course, intelligence could be created if organisms work in parallel, with each of their slower brains working in unison. A sort of aquatic hive might be envisaged.

Of particular interest is the discovery of complex, multi-cellular organisms living on the floor of the Mediterranean Sea. Roberto Danovaro and colleagues (Department of Marine Science, Polytechnic University of Marche) discovered a variety of simple multi-cellular organisms living in the sediments on the floor of the L'Atalante basin, a region filled with hypersaline (extremely dense and salty) water. These species, Locifera, were all new to science. Various techniques confirmed that these organisms were viable, metabolizing entities rather than strays that had simply fallen into the basin and died.

Unlike every other multi-cellular eukaryote thus far discovered, these Locifera did not run using oxygen. Instead their cellular mitochondria have been replaced with another structure (an organelle in biology-speak) called a hydrogenosome. Rather than consuming oxygen, these organelles produce hydrogen, hence the name. These hydrogenosomes utilize a similar strategy to mitochondria to generate ATP, but it is less intricate and generates less useful energy per unit of fuel than standard mitochondria. However, like mitochondria, they are clearly derived from free-living bacteria through the process of endosymbiosis (Chap. 7) and at least in some cases contain their own chromosome that carries instructions for the manufacture of proteins needed to make the hydrogenosome work.

Intriguing though these life-forms are, they suggest only that it is possible to evolve more complex organisms that operate without oxygen, not that you can evolve an energy-costly brain needed for intelligence. Oxygen appears essential, at least for the latter.

Finally, if organisms are to build devices capable of observing the universe and listening for the latest chart sensations from Earth – albeit 22 years later – they need dry land on which to construct them. An aquaplanet may be a hospitable abode for fish but is somewhat less likely a world on which to find the alien equivalent of Rosalind Franklyn or Richard Feynman.

Planet c, however, is not the final word in the habitability of the Gliese 667C system. Preliminary radial velocity measurements suggest the presence of an additional super-terran candidate (Gliese 667 Cd), orbiting in an *extended* habitable zone with a period of 91 days on a semi-major axis of 0.23 A.U. This would place it outside the tidal locking zone of Gliese 667C. At this orbital radius, Planet d may support liquid water, if large quantities of CO_2 and other greenhouse gases can maintain suitable temperatures. Gliese 667Cd may then resemble an overgrown version of Mars. Radial velocity data also suggest a final, Saturn-mass world (Gliese 667Ce) at an even greater distance, roughly equivalent to the Sun-Jupiter distance. This has an orbital period of several years.

Further observations are clearly warranted. Thus Gliese 667C may host two habitable worlds, mirroring Gliese 581. The inner Planet c, with its lower mass, most closely resembles Gliese 581g (if real), while Planet d may be a clone of Gliese 581d. Overall, this is an interesting astronomical twinning.

Looking further afield, the life of Gliese 667C will mirror that of its cousin, Gliese 581, and survive for hundreds of billions of years. As such, the driver of change on the planet will not be the slow rise in the luminosity of its star but the steady decline of its geological engine. Ironically, the slightly lower mass of Gliese 667Cc, compared to 581d, could well make its functional geological lifespan greater than that of the more massive 581d. This, too, makes 667Cc a far better prospect for a currently habitable world. Let's not forget, however, if Gliese 581g turns out to be a real world it may be even more suitable for life. Gliese 581g may contain a vibrant geological engine that still churns out carbon dioxide and recycles its dense oceanic crust.

The Development and Fate of Planet Gliese 667Cc

There are likely to be some differences in the development of planet Gliese 667Cc compared to Planet d of the Gliese 581 system. For one thing the formation of the three stars significantly alters the likely developmental plan of the Gliese 667 planets. The disc of material in which the three stars formed would have to have been much more extensive and more massive than that giving rise to Gliese 581. This would mean that larger planets would have been possible to manufacture. Indeed, this appears to be the case, with some of the more extended HARPS radial velocity data suggesting that a Saturn-mass world lies in an extended orbit.

However, we must not forget that star formation is a messy business, and there is always the chance that Gliese 581 and its developing worlds were originally part of a more extended family of stars, potentially part of a binary- or triple-star system, much like Gliese 667C is today. Gravitational interactions between protostars might have expelled the nascent Gliese 581 from the clutches of its siblings and off into a life of solitude. At present HARPS has ruled out any sizable additional worlds closer in than 5 A.U. in the Gliese 581 system, but there remains the possibility of further planets in a more extended family.

More pertinent, it is likely that the Gliese 667 system experienced a far greater period of planetesimal instability and scattering than was seen in the Gliese 581 system. This would have resulted from the more complex gravitational interplay between the three stars and the developing planets. In turn, this would have resulted in much more dramatic encounters early on in the life of the planetary system. It is more likely, therefore, that the Gliese 667 system experienced a more chaotic birth than the Gliese 581. Without orbital migration the worlds of Gliese 581 and Gliese 667 might have formed within the snow line and be relatively dry. However, given the greater opportunity for more chaos early on, the planets of Gliese 667C might have obtained a larger *secondary* supply of volatile materials. These would have arrived during a fierce bombardment of icy material as the stars and planets adjusted their orbits. Indeed the late heavy bombardment experienced by Earth might well pale into insignificance compared to the pummeling experienced by Gliese 667C's planets.

The age of Gliese 667 is uncertain but is probably considerably younger than Gliese 581. The only real clue to the age comes from stellar activity. The red dwarf Gliese 581 is a relatively placid star with low variability. However, Gliese 667C undergoes periodic flares. These temporarily bump up the level of X-ray and ultraviolet emission from the star. Flaring is very common in young red dwarfs and is a reflection of their rapid rotation and internal plumbing (Chap. 3). Flare activity tends to subside rapidly as the stellar magnetic field brakes against the interstellar medium. In most instances flaring is only a significant issue for stars less than 2–3 billion years of age. This implies that the stars of the Gliese 667 system are mere youngsters compared to the Sun and to Gliese 581. In turn this has implications for the type of life, if any, found on the surface, or in the oceans of Gliese 667Cc.

Despite relatively little to work with, we can still make some reasonable predictions regarding this most habitable of the planets, Gliese 667Cc. These require some leaps of faith, but they are not unreasonable. Soon they may also be testable.

We first assume that Planet Cc, with a relatively low mass of 3.4–6 Earths, is likely to be mostly made of rock and metal, with a variable amount of volatile materials, of which the majority is likely to be water. This is a simple reflection of stellar nucleosynthesis. Hydrogen is obviously extremely common, and oxygen is one of the most common "metals" (Chap. 7). Thus the surface is likely to be dominated by water; and at the temperatures predicted for Gliese 667Cc, this is likely to be mostly in the liquid and gaseous states, as it is on Earth. If greenhouse conditions are markedly more severe than found on Earth – perhaps because of a thick methane-rich atmosphere – temperatures could be sufficient to trigger a runaway greenhouse. However, with only 90 % the insolation found on Earth, and applying Occam's razor, this would seem relatively unlikely. We'll assume temperatures are similar to those found on Earth.

The proportion of land and water is very difficult to predict. A water-world may be probable if the proportion of volatile materials simply scales up with those found on Earth, and we assume the planet didn't experience any catastrophe early on that drove water out into space. If we assume that there is dry land, even in small, scattered areas, we can run a gedanken experiment of how the planet will evolve.

Recent studies of rocks from Mars reveal that these appear to contain the same proportion of water as those from Earth. This suggests a fairly straightforward scaling of volatile content with mass for rocky planets, at least within the planets of the Solar System. If we want Gliese 667Cc to have some dry surface it may well need to have been born drier than Earth if it is not to be flooded by the water vented from its hot mantle.

The planet forms 2 billion years ago from an extended and highly fragmented disc of gas and dust. Complex gravitational interactions between the three stars in the system scatters planetesimals in all directions and ensures that significant volumes of water and other volatile materials are available and contribute to planet formation. Three confirmed planets form a Saturnian body, first through direct collapse of part of the extended disc of gas around Gliese 667C or from the coalescing of a super-terran sized core that later accretes gas from the remaining disc. Indeed, formation of this world is completed within 1–10 million years, despite the relatively low masses of available material. Closer in to the star, large planetesimals collide within the snow line, forming relatively dry bodies. Orbital migration and planetesimal scattering allow the delivery of small icy planetesimals in through this line. These deliver water to these inner, dry worlds.

Meanwhile, the central star is still contracting towards the main sequence and has a few hundred million years to go until its core ignites. As it shrinks, the snow line migrates inwards until it approaches the orbits of the developing inner planets. James Kastings' work suggests that none of the habitable worlds in either the Gliese 581 or the 667C system would have experienced significant tidal heating. Therefore, we can probably disregard this as a factor in determining the amount of water either planet has. Thus the amount of water they have will reflects what they were originally born with and how much was delivered or removed during subsequent impacts.

Planetesimals are also being scattered in the direction of these inner worlds as Gliese 667C plows on through the debris leftover from the formation of the two main stars, Gliese 667 A and B. Much of this material is swept up by the developing planets. By the time the stellar core of Gliese 667C has ignited, each inner super-terran has accreted a substantial body of volatile materials, bequeathing each with an ocean of liquid water. Like the planets in our Solar System, the phase of accretion could be extended, depending on the dynamics of the star system and whether stars A and B have their own worlds. Planetary companions to A and B would also serve to scatter more material in all directions, some conceivably impacting on the worlds of Gliese 667C.

Beneath the waters of 667Cc, the motion of hot mantle material and cold, primitive crust is producing the first continental fragments. By 500 million years of age, the genesis of continents is underway. Land is tentatively poking its nose above the surface of the global ocean, and a vigorous cycle of carbon dioxide release and absorption begins. Over time the levels of this gas drop, but this is at a slower pace than the decline in the luminosity of the central star as it slips onto the main sequence.

With a more substantial luminosity early on, star C may have prevented Gliese 667Cc from being habitable. However, as the eons passed, the temperature fell and the star shrank.

This continued until the central star turned on its engines and stabilized. Any moist greenhouse collapsed with the declining burden of radiation, and the water, not lost to space, coalesced onto the planet's surface.

The steady decline in luminosity of red dwarf protostars is a very important and unique factor to consider when deliberating on planetary habitability. Protostars of red dwarfs spend a long time dimming while their planets steadily evolve. More massive K- and G-class stars all reach the main sequence long before terrestrial-size worlds form, but this is not true for terrestrial bodies orbiting red dwarfs. Thus Venus-like planets may be common in the early history of these systems, when the terrestrial worlds are blitzed by high levels of radiation from their protostars. Those planets endowed with insufficient reserves of water will lose it all during this phase. Water will be driven outwards into space by a combination of evaporation and the loss of hydrogen through photolysis. Photolysis is the splitting of water by light. Young red dwarf stars are proficient generators of ultraviolet and this is particularly efficient and breaking apart water to release hydrogen.

On Venus the planet's water has almost all been lost. The only clue to its earlier presence is the abundance of heavy hydrogen (deuterium) in the atmosphere. This isotope of hydrogen is present in excess, compared to expectations suggesting it is the fossil footprint of former oceans, long since boiled away into space.

However, if a planet can survive the early roasting, or if more water is delivered to it by an analogous late heavy bombardment, those endowed with sufficient water become more Earth-like as their temperatures fall. Indeed, losing water to space in an early wet greenhouse phase may make these planets *more* habitable for complex species, not less. If super-terrans are initially endowed with too much water, loss of much of this in an early moist greenhouse may make it more likely that these worlds could have areas of continent as well as ocean. A shallower ocean would allow land to emerge and an efficient carbonate-silicate cycle to operate.

What next? Our super-terran resides in the habitable zone of its star and has a shallow, yet pervasive, ocean covering its surface. Beneath the waters the hot silicate mantle generates a dense komatite-rich crust, but also fills the early oceans with hydrogen sulfide and ferrous iron. These maintain a thick, anoxic (oxygen-free) environment under which current terrestrial life would suffocate. Unpleasant though this sounds, this rich brew of chemicals is ideal when it comes to producing biomolecules. With a planet perhaps twice the diameter of Earth, the global ocean contains quadrillions upon quadrillions of opportunities for experimentation.

If Earth is anything to go by, molecules combine in ways that allow error-prone molecular replication soon after conditions become conducive. This is an inexact process of duplication. Crystals of minerals, such as olivine, reproduce by the simple addition of silicate, magnesium and iron ions to build longer and longer duplicated molecules. Although some "mutations" occur in which other metal ions slip in by mistake, this is fairly rare. In biological systems, the process of copying molecules is sloppier, and more mistakes are made. This may seem problematic, but it is actually a beneficial process, where trial and error builds progressively better molecular machines. As with human behavior, life learns from its mistakes. Error is the bedrock of evolution through natural selection.

Over time a diverse brew of different molecules arises, some of which are competent at copying themselves; while others are less able to, or utterly deficient in the process. At each stage the probability of success per molecule is very low, but with such vast numbers of molecules available, and with time spans running into millions of years, the overall probability of success is high, verging on the inevitable. There are quadrillions upon quintillions of opportunities – numbers vastly more than our primitive brains can readily comprehend.

What molecules are the genetics of life on an alien world based on? Our anthropocentric sensibilities tend to enthuse about deoxyribonucleic acid (DNA) or its more malleable sister molecule, ribonucleic acid (RNA) as the bastions of life in the universe (Chap. 7). However, there is absolutely no reason to believe alien life would use these as its store of information.

From infrared observations of interstellar clouds we can say that amino acids are everywhere, and hence proteins will probably be universal. These molecules are found throughout the cosmos, from carbonaceous meteorites and asteroids to the star-stuff of nebulae. Given their ubiquity, amino acids are more than likely found functioning in some regard in all living organisms, regardless of the location. Simple sugars, alcohols and simple organic acids are also common in interstellar space; thus we can reasonably assume they will also play a part in the biochemistry of aliens.

However, DNA and RNA are built from three core components: a phosphate ion, a five-carbon sugar and one of four nitrogen-rich compounds called bases (Chap. 7). Phosphate and possibly adenine excepted, none of these other building blocks have definitively been found in interstellar space. The sugar ribose could be replaced with simpler alcohols such as glycerol, which is present in extraterrestrial environments such as nebulae. Likewise, the bases can, in part, be assembled from cyanide, which is also ubiquitous in nebulae, comets and carbonaceous meteorites. However, whether the organic chemistry or primitive biochemistry would follow the same path on another world is anyone's guess. It certainly would seem improbable. Panspermia aside, if we discover life on Europa, Enceladus and Mars carrying the same genetic instructions as ours, it will be an amazing find.

If we look at Earth, life appears to have arisen shortly after the late heavy bombardment ended, some 3.8 billion years ago. Examination of the Moon suggests that this was a brutal phase that may have eliminated early life or prevented its rise. However, the late bombardment might have played a more critical part in the rise of life than we imagine, delivering the essential ingredients of life on impacting comets and asteroids. That aside, a key issue here is the apparent rapidity of life's rise following what were undoubtedly very difficult times. Although our sample is only one, the rapid emergence of life suggests that it is a highly probable event on a planet – any planet. This clearly ties in with the assertions made above regarding the probability of reproducing biomolecules arising in any ocean. Thus, we can reasonably expect life to arise on any world where conditions are conducive. In geological terms the rise time is likely to be fast, but will clearly be determined by the local chemistry of the planet and perhaps the nature of the objects that impact upon it.

So, all things being equal – admittedly a somewhat questionable position – we can expect that Gliese 667Cc has life, if its surface is similar to ours and at a comparable age. A deep ocean and thick, opaque atmosphere may deter or prevent the rise of life, but this is less clear (Chap. 9). What we do know is that life can arise where there is water and suitable but diverse chemistry. With an ocean more than 20 km in depth, there won't be any solid surface, but life can readily arise and utilize geothermal energy supplies. Later some of this could evolve the characteristics needed to absorb sunlight, giving rise to a more Earthly biosphere.

If the atmosphere is dense and little light or stellar heat permeates it then the only life that will be possible will be that driven by internal chemistry and heat. Under these circumstances it is unlikely that oxygenic photosynthesis will arise. The available geothermal sources have too little energy to drive photolysis (Chap. 8), and the abundance of volcanic hydrogen sulfide would most probably retard the evolutionary drive needed to instigate the process of oxygen production. Quite simply, given an abundance of one resource there is no selection pressure to drive the acquisition of another. Think of it this way: when oil was plentiful and cheap you'd have been hard-pressed to find many automobile companies willing to invest in engines driven by an alternative fuel. But now that oil is a pricey and sought-after commodity, car manufacturers are scrambling to provide electric and dual-fuel engines. Similarly, an oceanic Gliese 667Cc may well host life, but it will most likely be microbial and anoxic – intriguing, yes, but not likely to build another Empire State Building.

If the oceans are less than 10 km deep, there is the potential for plate tectonics and volcanism to build land above its surface. The deeper the ocean, the less the land will be aerial – exposed to the atmosphere. Moreover, with a larger planetary mass, Gliese 667Cc will have a greater surface gravity. This will encourage the spreading of any continental crust under its own weight, lowering its height. Consequently, there will be less scope for either plate tectonics or mantle plumes to lift continental crust above the surface of any ocean.

Thus, ideally, we want Gliese 667Cc to have relatively little water – perhaps equivalent to that of Earth or only marginally more so. We should, however, be aware that if the proportion of volatile materials scales with planetary mass then the bulk of water available to form oceans will be greater and the depth of any oceans more. Our habitable world – habitable for complex, land-dwelling life, will, therefore, have to be proportionately drier than Earth for its greater mass.

In our scenario, Gliese 667Cc has less water, and the land is able to rise above the surface of its oceans once continental crust forms. Populations of islands, both hot-spot and plate tectonics related, emerge by a few hundred million years at most after the planet forms. The atmosphere, at this point, is still dominated by water vapor, carbon dioxide and nitrogen. Smaller quantities of methane are produced by volcanic processes, occurring on the ocean floor and by any emergent anaerobic life. Some is also supplied by the steady, but declining rain of comets and asteroids.

Running forward, chemical processes of the types discussed in Chaps. 7 and 8 also begin the process of capturing light energy. This will only happen if the atmosphere is suitably transparent to visible and near-infrared (shorter and more energetic infrared) wavelengths. Although life has emerged in the ocean depths it can adapt through natural selection to incorporate the incoming energy of the parent red dwarf star.

On Earth there is evidence of anoxogenic photosynthesis as early as 3.5 billion years ago in the form of isotopic signatures and, more contentiously, the presence of organic compounds found in the photosynthetic membranes of the now fossilized microbes. Examination of the tree of life – a branched pictogram of all living organisms based upon their DNA sequences – indicates great antiquity for anoxogenic (oxygen-free) photosynthesis. It also suggests a hot origin for life – perhaps emerging around hot springs or undersea volcanic vents. However, once the available supply of nutrients needed for growth and development runs low, there is obvious selection pressure for the development of processes that produce these materials rather than merely consume them from their surroundings. There is a gap of nearly 300 million years between isotopic evidence for life and fossil evidence for photosynthesis, suggesting that life was either making do with what nature had provided, or that the evidence for earlier photosynthesis was destroyed by Earthly processes.

Approximately 300 million years after anoxogenic photosynthesis emerged, fossil evidence points to the formation of rocks called kerrogenous shales. These rocks show that, at least locally, oxygen was being manufactured by microbes. There is additional evidence, in the form of the wonderful structures called stromatolites. These structures are built from the dead remains of bacteria and mineral ions extracted from seawater. Stromatolites are produced by diverse communities of bacteria, organized around cyanobacterial factories. On Earth it is the process of photosynthesis occurring within these bacteria that produces oxygen in the greatest abundance by photosynthesis. The cyanobacteria lurking within stromatolites are the ancestors of the photosynthesizing chloroplasts of plants. This biological transformation has had the greatest impact on life on our planet – and the biggest impact on the geosphere as a whole. As the source of oxygen gas, these organisms rusted our planet and ultimately led to the formation of our unique oxygen-rich atmosphere (Chap. 7).

Within Earth's oceans, oxygen rapidly reacted with dissolved metals and metal sulfides, particularly compounds of iron and sulfur. On Earth, the level of atmospheric oxygen stayed at less than 1 % for 2 billion years. The key to the delay between the appearance of atmospheric oxygen and the emergence of photosynthesis was likely due to the effect of materials dissolved in the oceans. A super-terran, with a voluminous ocean, will take longer to oxidize as photosynthesis pumps out oxygen. Thus, although oxygenic photosynthesis may arise early, it will take longer for this gas to permeate first the oceans and later the atmosphere. Moreover, while this delay is in place microorganisms and volcanic processes will have more time to flood the atmosphere with methane. Thus the atmosphere of a super-terran may well take longer to become oxygen-rich despite a comparable initial history.

If we assume that Gliese 667Cc is less than 2 billion years, based on the relatively frequent rate of flaring in Gliese 667C, then our habitable world may have a very unpalatable atmosphere that is free of oxygen gas at present. We may have to wait a further 2–3 billion years before the atmosphere becomes habitable for humans, or indeed the majority of complex life from our world. If we are looking for worlds habitable for complex oxygen-breathing life then we may be looking in the wrong place. A less massive, more Earth-like world may be a more promising prospect than a super-terran.

Of course, these are fairly anthropocentric assumptions. A larger planet may well have more methane and more sinks for oxygen. However, it may also have more sinks for these oxygen-loving materials. Conversely, a greater bulk of oxygen-generating cells in the wider, and perhaps deeper, oceans of the super-terran may produce *more* oxygen, and hence the atmosphere may fill more quickly. If microbes take it upon themselves to use methane instead of carbon dioxide as their carbon source, then the atmosphere could be fairly light in terms of this gas. Consequently, methane may not impose such a restriction on oxygen's rise.

What of planetary climate? Temperatures will fall over the first 500 million years or so of the planet's life as the central star composes itself and settles onto the main sequence. Tidal locking will ensure that one face is permanently lit while the other dark. The sunlit side of Gliese 667Cc is predicted to have temperatures well above the melting point of water ice – possibly as high as 30–40 °C, making its temperature similar to Earth in 1 billion years. Atmospheric models repeatedly show that despite permanent darkness, the shaded side will maintain temperatures at above –20 °C as long as the atmosphere is thick enough to transport energy from the lit side. For an Earth-mass planet or the super-terran, Gliese 667Cc, it is more than reasonable to assume that it has an atmosphere thick enough to transport heat efficiently. As we've already seen, the proportion of volatiles in mantle rocks from Mars and Earth is the same despite gross differences in planetary mass. Therefore, our proportionately more massive super-terran is likely to have a generous atmosphere enriched by gases from its hot mantle.

If Gliese 667Cc is an aquaplanet, its atmospheric circulation is likely to be fairly simple. Assuming that the planet rotates once on its axis per orbit, the Coriolis effect will be weak. The spin of Gliese 667Cc is 28 times slower than on Earth. Modeling suggests strong movement of air across longitudes and through the terminator (the line separating light and dark sides). Where heating is strongest, air rises by convection, generating areas of low pressure. The coldest regions are likely dominated by high pressure, where cold air is sinking and spreading outwards towards the warmer, sunlit side. However, as Chap. 6 illustrated, it is unlikely to be a simple arrangement, with high pressure on the cold side and low pressure on the hot side. The circulatory pattern will be more complex, in part because of the planet's spin but also as it is unlikely that the warm, rising air will be buoyant enough to flow right around the planet at great heights. Although weak, the Coriolis effect will also provide some east–west deflection of moving air. Therefore the pattern of airflow will be more fragmented.

On a planet with a continent or two, the movement of air will be even more complex. Areas of land warm more easily than areas of ocean. Air will tend to rise most strongly over sunlit land (Chap. 6). Thus any sunlit-driven convection will be focused over illuminated land. Mountains – orography – will also have profound effects on air flow and temperature. The illuminated side of mountains will face Gliese 667C more directly than flat land at a similar latitude and longitude and thus will warm more. This heating effect, coupled to the simple blocking action of mountainous on airflow will lead to the generation of low pressure on the sunlit side – somewhat akin to the formation of the Tibetan low in the summer of Earth's northern hemisphere. In the cold, permanently shaded lea of the mountains, cold air will slump downward, generating areas of higher pressure. Meanwhile, around the margins of any cold continental areas, warmer oceans will tend to focus areas of lower pressure.

Airflow will also be deflected by any landmasses and high topographic features, and as Chap. 6 showed, the spin of the planet is likely to produce super-rotation within the atmosphere. This will generate a belt of air flow, either at the surface or high up, from the warm, sunlit side to the cold, dark side. This flow is most likely to occur along the equatorial regions, transporting warm air east, while cold air predominantly flows from the dark side, across polar regions towards the warmer hemisphere. Thus Fig. 10.3 is purely hypothetical and most likely grossly simplifies the flow of air, even for a world with a stable climate.

On Earth the atmosphere is dominated by oxygen. High up, this gas is altered by ultraviolet light to form ozone. Aside from protecting the surface of the planet from harmful UV rays, the absorption of ultraviolet light also warms this layer, forming a stable, stratified zone we call the stratosphere. This warm layer forms a lid, preventing the vertical movement of air from the surface to extreme heights. Consequently, the coldest atmospheric temperatures are found not at greatest heights but nearer the surface. On Earth convection of warm air over tropical areas is

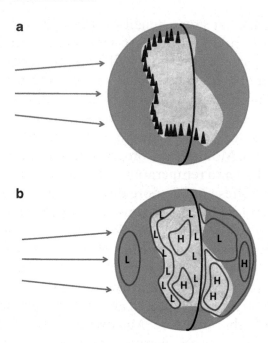

FIG. 10.3 The effect of topography on the distribution of winds on a tidally locked planet. A large supercontinent in (a) occupies the center of the disc with mountains fringing the sunward side. Sunlight is intercepted by the mountains, causing greater heating on the lit side, but stronger cooling on the shaded side than would be expected for a flat world. In (b) this generates areas of convection (low pressure, L) on the warm, lit side, but areas of downwelling (high pressure, H) on the cold side. In addition ocean circulation around this continent will redistribute heat in unexpected ways

deflected north and south after rising only 30 km or so. On a planet with no effective absorption of radiation, or one where the upper atmosphere is more opaque to incoming radiation, this lid will be higher. Air can thus rise further – assuming there is sufficient energy to drive its upward flow.

Like all tidally locked worlds, the chance of Gliese 667Cc supporting even one sizable moon is unlikely. Tidal forces between the planet and the Sun should make any orbits unstable, disassociating any moon from its parent world. However, there is no need for a moon to stabilize the planet's inclination. Tidal forces with the parent star should do the job alone.

Future World

Let's extend our model from the present, assuming that Gliese 667Cc is currently 2 billion years old. Given appropriate opportunities an oxygen-rich atmosphere arises in the future, driven by the evolutionary acquisition of oxygenic photosynthesis. What then?

With land on which to colonize, the oxygen gas reacts with available ultraviolet light – something that is less abundant radiated from red dwarfs – forming a protective ozone screen. Life can then emerge onto the land and begin to diversify. What we know from our experiences here, given opportunities, life will fill the biological gaps. Complex life is a near certainty given an oxygen-rich atmosphere, time, and a diverse range of environments on which to colonize. Whether *intelligent* life is a certainty is another matter. We really have nothing to go on except our own experiences on this planet. Rudimentary intelligence certainly is common on Earth, but it has taken the best part of 4.6 billion years for life to compose a book using a computer. We simply don't know.

On Earth the reasons for intelligence arising are unclear. Although we know that octopi are intelligent – perhaps comparable with a dog – they appear to have devoted the vast majority of their brain capacity to using visual information. This will have been driven by the need to survive in a shallow oceanic environment where there is strong variation in light and color throughout the day. An octopus can change its color and shape to blend easily in with different backgrounds. In turn this allows it to evade predators and catch prey.

Whereas invertebrates have primarily low intelligence, mammals are predominantly high intelligence organisms. There is considerable variation, but on the while they are more intelligent than their reptilian cousins. The question is why. Almost certainly, this will relate to endothermy – the ability to maintain a warm body, independent of external conditions. With a warmer body metabolic activity will be greater, and there will be more consistency in this throughout the day. Another interesting idea is that mammalian intelligence arises because animals are furry, tactile and need to cooperate more consistently in social groups. Furriness may seem like a superfluous point, but mammals appear to like being stroked. They like having their hair brushed, and they like being touched. Their brains then provide a chemical reward that makes them feel good about this – dopamine. As such, tactility, which appears to have followed the evolution of hairiness, promotes socialization. Life within a group demands more brain power to cope and manage the profound increase in the amount of interaction. Perhaps, then, having fur is what a organism needs to become truly intelligent. If this is true, then the requirement for the development of intelligence is the presence of dry land on which to evolve warm-bloodedness (endothermy), and tactility then follows from the desire to touch. Tactility and social interaction then leads to a larger brain.

Let's once more assume that complex life can arise if there is dry land and that intelligence may then arise if the environment allows the evolution of warm-bloodedness. What then are the consequences and likely fate of such life given the geological constraints of life on a super-terran? If the planet has a mass between 1 and 4 Earths then the expected geological processes will eventually clog the surface with granite and plate tectonics will cease, irrespective of the amount of internal heat available to drive it. Hot-spot volcanism could then build increasingly large surface structures, much as it has done on the stagnant surface of Mars. However, the height of these will be limited by the constant undermining flow of material at the base of the crust and the plasticity of the underlying mantle rock (Chap. 5). Given enough time – perhaps 6 billion years or so after birth – the land may be unable to raise its head above the surface of any oceans and will disappear from view.

As the land erodes and slumps back down into the abyss, the carbonate-silicate cycle will grind to a halt (Chap. 5). Not only will the loss of land mean a progressive and protracted phase of global extinction, but it will also mean the failure of the global thermostat. What happens next sensitively depends on the amount of carbon dioxide remaining in the atmosphere and the change in this level over the ensuing billions of years.

On Earth the steady evolution of our star has meant slowly increasing surface temperatures. This has been offset by increased rates in the hydrological cycle that generates rainfall and other precipitation. Higher cycling of water between oceans and atmosphere has drawn more carbon dioxide from the air, causing the gas to react with silicate rocks. This process in turn lowers the global greenhouse effect. This restores temperatures to within a habitable mean (Chap. 5).

On a planet without dry land there are no silicate rocks to erode, and such an effect is removed. Carbon dioxide levels are thus balanced between the atmosphere and oceans in a manner dictated solely by water temperatures. At higher ocean temperatures carbon dioxide is less soluble and builds up in the air. At lower temperatures the converse is true, and lower temperatures mean less carbon dioxide in the air. On a super-terran or other world without land to erode, atmospheric levels of carbon dioxide – and other gases such as methane – will be dictated by the following processes: the production of carbon dioxide by undersea volcanic vents; the temperature of the ocean; consumption and production by photosynthesis and respiration; the loss of oxygen to space; and ultimately most significantly the painfully slow, inexorable rise in stellar luminosity.

Thus an interesting scenario arises. Life emerges with equal rapidity on these worlds as it did on Earth. Life then follows a similar path, dictated in large part by geology. However, as the land wanes, life retreats back into the oceans. For billions upon billions of years life remains confined to the oceans at a level dictated by the biological productivity of the carbon dioxide dissolved in the water. Once the geological supply of carbon dioxide, and or methane, wanes life contracts and eventually becomes extinct. Hundreds of billions of years later, when Gliese 667C evolves further, the oceans begin to boil away, exposing the thick, fossilized remains of a once vibrant community.

The course of life on a super-terran might then follow an odd path, confined to a narrow geological window. Life spends perhaps 4 billion years emerging and rising in its complexity, much as it did on Earth, only to fall another 1–3 billion years further on when the land erodes back into the oceans. During its heyday, complex plant and animal life forms visibly dominate the land and oceans. Out of sight, the true masters, the microbes, dominate the planet as a whole. As the land erodes, one by one, complex species become extinct. Complex and microbial ocean life continues relatively unscathed. For perhaps another few billion years hence, life remains in the oceans, but with slowly declining levels of carbon dioxide it retreats further. By 10 billion years or so, all complex life that hasn't adapted to the declining rates of carbon dioxide production goes to the wall, leaving the microbes once more to run planetary affairs unaided. Conceivably, with even low rates of carbon dioxide released from the interior, microbial life could continue almost indefinitely. However, life would undoubtedly find it difficult to keep going throughout the full 400 billion years until the central sun burns out.

At the end of the star's life on the main sequence, the oceans are evaporated. There may exist a brief window when the evaporating waters give up a little carbon dioxide from the underlying rocks, before temperatures become too high for even hardy microbial life. Life might have one more go before the waters are lost forever, but it will be a brief stint.

The Geodynamo and the Preservation of the Atmosphere

As we have seen it appears that plate tectonics is needed to maintain the global geodynamo on Earth (Chap. 5). To recap, without the cooling effects of subducted oceanic crust upon the hot core, the circulation within the liquid iron may become too sluggish, causing the geodynamo to fail. This assumes that a slowly rotating planet can even support an active dynamo. Mercury is our only working example of this kind of world, and Mercury may not scale up to a tidally locked super-terran. Venus also rotates slowly and has no magnetic field. Small Mercury may sustain one because it has a steeper temperature gradient across its small core. Conversely, the bulkier Venus simply doesn't have a suitable temperature gradient, being hotter throughout. The combination of a slow spin and small temperature gradient makes a geodynamo impossible on this larger world. If it is Venus rather than Mercury we need to scale up then the habitable planets of red dwarfs may universally lack protective magnetic fields, even where plate tectonics still operates.

This has important ramifications for any planet orbiting a red dwarf. If plate tectonics is naturally limited by planetary heat death, or by the formation of a thick lid after 5–6 billion years, this period of time might be the limit to the lifespan of a geodynamo.

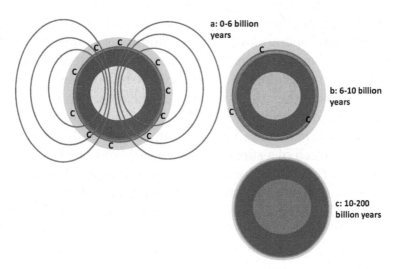

FIG. 10.4 The decline and fall of a super-terran. In (a) the young planet nurtures a dense atmosphere infused with carbon-dioxide, "C," that supports photosynthesis. In (b) plate tectonics has ended and the geodynamo is dead. The atmosphere loses the bulk of its carbon dioxide through geological processes, terminating photosynthesis, at least on land. Without a protective magnetic shield, an old super-Earth (c) has lost most of its atmosphere to space, and its former deep oceans have largely evaporated. Atmosphere and other layers are not drawn to scale

What happens once the field fails? Quite apart from providing some protection for any surface life forms from the harmful, mutagenic effects of cosmic rays, the protective magnetized barrier blocks much of the action of the stellar wind on the planetary atmosphere. Without a magnetic field the stellar wind interacts directly with the atmosphere and begins to erode it.

We can see the effect of the solar wind on the atmosphere of Venus. Although the varying topography and evidence for granite suggests Venus once had plate tectonics, it certainly does not operate now. Venus may have been a viable planet for life, with oceans that provided the water to drive active plate tectonics, until a billion years ago. However, once the planet became stranded on the hot side of the habitable zone, its oceans evaporated and plate tectonics ended. Nowadays, without a protective field, the Sun's buffeting wind is stripping oxygen and other gases out into space. So extensive is this trail of atmospheric debris that some of it reaches our world millions of kilometers away (Fig. 10.4).

Although the stellar winds of red dwarfs are predominantly weak, they will operate over tens to hundreds of billions of years and must have measurable effects even on super-terrans with initially thick atmospheres. If, after several billion years, the magnetic field falters the atmosphere will be exposed to the effects of the stellar wind. Day by day, ton by 100 t, the atmosphere will be chipped away. Depending on the composition and density of the atmosphere, much of a super-terran's water could be removed over the course of the life of the star. Thus a water-world may steadily dry out as it orbits its persistent central star. If life can cling on for most of this period, the eventual emergence of land on water-worlds might finally allow a burst of diversification before all the carbon dioxide is once more exhausted and the waters evaporate completely.

We could consider another possibility: the geodynamo persists, but given the proximity of the planet to its star, it is directly coupled to it. Such a pairing might inflict serious damage to the atmosphere of the planet by directing coronal mass ejections and other explosive outbursts to the planet. If this occurred then the planet's atmosphere might be eroded even faster. Again, this is speculation, but observations of stellar activity might allow this possibility to be investigated.

The most important conclusion is that the proportion of the planet's life that is habitable is far less than the lifespan of the central star. The only long-term hope for life is that intelligence evolves and can then use its brain power to adapt to the deteriorating conditions. With intelligence, life could move out and colonize the oceans and still have the capacity to leave the planet before conditions become adverse for higher life. Geoengineering would be a must if complex life is to maintain habitability throughout the life of its star. If an Earth-mass world is substituted for a super-terran, conditions will deteriorate over a similar geological period. Where the planet has lower mass the cessation of plate tectonics is driven by the heat death of the planet. Once plate tectonics ends, the same loss of carbon dioxide ensues and life is condemned to the same fate (Fig. 10.5).

With more massive stellar hosts, planets either lose carbon dioxide at a pace similar to stellar evolution (G-class stars), or its loss is outpaced by stellar evolution (F-class stars and earlier).

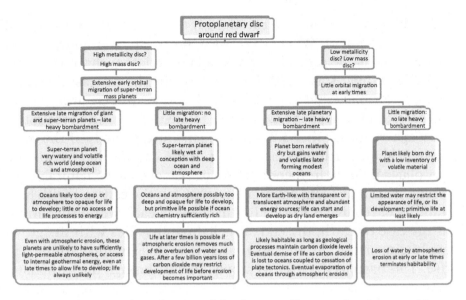

FIG. 10.5 The interplay of some of the factors affecting the short- and long-term habitability of super-terran and terran planets around red dwarf stars. The initial inventory of water is perhaps the most important in the short term, with the loss of the geodynamo and atmospheric erosion playing the critical role at later times. Ultimately, the availability of water and carbon dioxide will determine the long-term habitability of red dwarf worlds

Either way habitability is likely limited to less than 15 billion years on planets without some form of geoengineering. Modeling work by Werner von Bloh and colleagues indicates that where the surface of the planet is dominated but not totally covered in water a planet with a mass of a little over three times Earth may be habitable for up to 15.4 billion years. This figure extends to 16.3 billion years where the mass is 4.3 Earths. However, in all models this clearly depends on how much of that mass is water and other volatile substances. Increase these to the point that the surface is completely flooded and there will be no conventional silicate-carbonate cycle.

Remember also that water-encased worlds, or where land is negligible in its area, are unlikely to produce organisms that are capable of developing the intellect or technology required to escape their world or to communicate with the outside universe. Thus, regardless of the initial mass of the planet, assuming it has land on which life can develop, intelligence will be required to maintain habitability in the longer term – longer than the planet's geological engine can support. Where developed, however, habitability could last as long as the central star, given appropriate technology.

For comparison, Earth has been habitable, at least in terms of the solar habitable zone, for 4.6 billion years. Assuming nothing catastrophic happens, Earth will retain oceans and a breathable atmosphere for another 1.2 billion years. Interestingly, if the magnetic field on Earth fails, as a result of the cooling and freezing of our planet's liquid iron core, then our planet may be habitable for longer. This may seem counterintuitive, but the contrast with habitability of red dwarf worlds lies in the way each world succumbs. In the case of Earth, the end of its habitability lies with the growing luminosity of the Sun. As it brightens global temperatures will rise. This hastens the loss of carbon dioxide through the carbonate-silicate cycle, but also, ultimately, causes the oceans to evaporate. Without water, all life ends in a torrid, humid greenhouse.

If, however, Earth has lost its magnetic field at 1 billion years, or it is weak, the solar wind can begin stripping gas from our atmosphere. As nitrogen and oxygen are blown away the density of gases in the atmosphere will go down. This will weaken the greenhouse effect indirectly. Nitrogen and oxygen gas molecules normally bump into carbon dioxide and water vapor molecules in the air. This broadens the range of infrared wavelengths they can absorb. With less of these gases in the air, the ability of carbon dioxide and water vapor to absorb and retain heat is lessened, and Earth will heat up more slowly with the brightening Sun. Calculations show that if enough nitrogen and oxygen can be blown away and the steady loss of atmosphere can keep pace with the growing brightness of the Sun, Earth can remain habitable for another 3 billion years. Over this time the oceans gradually evaporate, but sufficient water should be left to maintain habitability over this extended period of time.

Although the rise of the wet-then-dry greenhouse is the more likely scenario for the future Earth, if enough gases can be lost from the atmosphere it would mirror the ultimate fate of a super-terran planet orbiting a red dwarf. These, too, must eventually succumb to the effects of stellar wind, rather than the rise of a super-greenhouse. It will be the loss of the atmosphere and oceans that will ultimately end their habitability.

Perhaps then the most important consideration regarding the habitability of red dwarfs will be the fate of their internal dynamo. How quickly after birth will these tidally locked super-terrans lose their magnetic fields, if indeed they are ever born with one? If tidal locking shuts down the geodynamo early, then these worlds may become uninhabitable even earlier than ours, perhaps within a couple of billion years depending on the thickness of their atmosphere. At present we can only speculate. If tidally locked worlds lose their magnetic shield early, this would be a thoroughly miserable prospect for habitability, given the ubiquity of red dwarfs and the almost unimaginable period of time they persist. Even with a gloomy scenario a planet born with a sufficiently dense atmosphere will retain an atmosphere for billions of years. The key is whether what they retain by the time life arises is sufficient to nurture its development. After that life will require intelligence to either manage the loss of gases in some manner or to leave and find somewhere else more amenable to its persistence.

Stellar Evolution and the Ultimate Fate of the System

Beyond the end of Gliese 667Cc as a habitable world, we must examine the fate of this system. Ultimately, the death of the star system may lie with its two orange K-dwarf stars. Not only will these evolve on a timescale more aligned with the life cycle of the planet, but their ultimate demise as white dwarfs will have important ramifications for Gliese 667Cc, even though it orbits the smaller, more distant star. Thus it is to the death of two K-dwarfs that we will first turn our attention.

Throughout their main sequence, or hydrogen-burning, stage, they will slowly brighten as their cores fill with helium. As each does so it will contract and grows denser. As far as the distantly orbiting Gliese 667Cc goes, these more subtle changes will have no measurable effect. An observer on the surface of our habitable world, or rather billions of generations of observers, would merely perceive a slow rise in the brightness of its night sky. Surface temperatures would not change in any measurable way.

After approximately 20 billion years, star Gliese 667A will leave the main sequence, and over the course of a few billion years swell into a red giant. The process will be more protracted than experienced by our star, as it is driven by the steadily increasing mass of the stellar core. In turn, this rate is driven by voracity of the hydrogen-burning shell that surrounds it (Chap. 3). Within each star this will run more slowly than it will in ours, as the core temperature is initially lower. During this first ascent of the red giant branch, the newly born red giant will finally shed a more glamorous light on Gliese 667Cc. Although not overheating the distant planet, at nearly 1,000 times its original luminosity, Gliese 667A will be bright enough to dissipate the near eternal night on the tidally locked planet's night side.

As planet Gliese 667Cc orbits its red dwarf Sun, alternate sides of the planet will face the red giant. For part of the planet's 28-day orbit, both the distant red giant and the planet's red dwarf will light up the sky, with the red giant rising and setting in a semblance of the day-night cycle experienced by freely rotating worlds. As the red giant sets on the day side of the planet it will illuminate the night side. Thus for several billion years the expanding red giant will ensure the entire planet is illuminated for at least some of its short year. Whether there will be any evolutionary consequences is harder to determine. Certainly, for an extended part of the planetary orbit night will be banished temporarily, and the entire planet will be lit. However, the amount of light received from the giant may be too little to allow for the growth of photosynthetic organisms – assuming, of course, sufficient carbon dioxide even remains in the atmosphere at this stage (Chap. 5).

During a brief phase, lasting no more than 150 million years, Gliese 667A will burn its helium ash. Thereafter, a second, transitory red giant phase will terminate with the expulsion of the star's outer layers as a planetary nebula. The shrunken core will endure as a dwindling white dwarf, its pallid, silvery light lost in the depths of the star system.

The passage of the outer layers of the star will likely have no direct consequence for Planet c, other than producing rather spectacular displays, perhaps even stunning auroras. However, during each red giant phase, and the final planetary nebula stage, there is a substantial loss of mass from the star system. This will increase the orbital separation of star C from stars A and B. This will also cause star B to drift away from the diminishing A. Star C will thus be more prone to escape from the grips of the other two stars if the star system interacts with other celestial bodies. When the sun evolves into a red giant and finally a white dwarf, the surviving planets will be left orbiting a star with 55 % its current mass. The white dwarf remnant of Gliese 667A will weigh in at roughly 0.51 solar masses, its mass having gone down by a third.

Perhaps 20 billion years after star A completes its evolution, star B will follow suit. After a similarly spectacular, but relatively brief interval basking in the light of two suns, Planet Cc will once again return to the normality provided by its single sun. The red giant phase of Gliese 667B will have a further interesting consequence for the star system as a whole. Although not a tight partnership the shriveled but enduring white dwarf remnant of Gliese 667A will almost certainly accrete some of the red giant's wind. Although not likely a regular occurrence, if sufficient hydrogen is grabbed by Gliese 667A it might just fire off a few nova outbursts. A nova occurs when a white dwarf with a mass less than that of the sun accretes roughly one Earth-mass of hydrogen from a companion star. The hydrogen compresses and ignites, undergoing violent and rapid conversion to helium.

Particles of antimatter (positrons) are released as a by-product of these fusion reactions and cause dramatic heating of the thin hydrogen shell. The entire shell, plus some of the underlying white dwarf, is catapulted into space in a matter of minutes, causing a brilliant increase in the luminosity of the white dwarf. If Gliese 667A and B are close enough when star B becomes a red giant, the ancient worlds of Gliese 667C could be in for a spectacular treat.

During the demise of stars A and B, star C will hardly have changed. Its photosphere will be slightly hotter, the star will be somewhat brighter and its light a little bluer. However, star B, having completed its red giant phase, will leave even less mass in the system. Star C will then be compelled to orbit even further away from the system's center of mass.

Perhaps billions of years later, star C and its worlds might suffer a fatal gravitational encounter with another passing star and be dragged away from its moribund compatriots. Planet Cc will be left with a single sun; its former stellar partners a distant and fading memory.

Sustainability of Life on Other Worlds

We've explored the rise and fall of life on super-terrans. The processes will most likely be identical to those on smaller worlds and will be subject to an eventual fall, dictated by the failure of the planetary engine rather than the central star. Can a planet remain viable for longer than 10–12 billion years? It would seem geology always deals the winning hand. However, let's assume we have intelligent little green men and women. What can they do to maintain the habitability of their planet?

If intelligent life arises and can produce the kind of technology capable of engineering its world, then some options emerge that increase the endurance of life overall. Without much evolutionary drive from the central star, the only variable worth considering is the sustenance of carbon dioxide level, necessary for photosynthesis. This is dependent on either the sustenance of a silicate-carbonate cycle and the geodynamo. On a planet with declining carbon dioxide, carbon will primarily become locked in carbonate rocks – mostly calcium and magnesium carbonate. Carbon dioxide is readily liberated from these carbonates by heating them to a few hundred degrees. Assuming the occupants of any such world have long since exhausted their quotient of fossil fuels – something we are adept at doing regardless of the consequences – then they can turn to their massive store of carbonate rocks. These will have built up over the course of the planet's evolution as carbon dioxide in the air reacted with water and silicate rocks.

Using abundant solar energy (or nuclear energy) furnaces could cook the carbon dioxide out of these rocks and sustain atmospheric carbon dioxide at levels suitable for photosynthesis. Alternatively, they could import hydrocarbons from neighboring worlds and combust them. The latter would be a more costly solution but not unreasonable. At this stage, active management of carbon dioxide levels might sustain habitability for tens of billions of years. The next obstacle would be the fall of continents…

Without the geological processes needed to keep the continental head above the surface of the waters a combination of erosion and the steady, albeit slow flow of continental crust over the surface of the mantle would mean that land slowly drowns and disappears from view. Any intelligent life would need to adapt to this change. Geoengineering might be an option – if they chose to evaporate their oceans to keep pace with the change. However, the consequences on life in general would likely be severe. Splitting water to liberate hydrogen and oxygen could also be done, but the scale of such ventures would need to be immense and likely prohibitive. Instead, adapting to a life on the ocean waves would seem probable. Constructing floating ocean cities, or migrating to live under the ocean surface would seem inevitable – assuming they choose not to leave their world altogether.

The time spans of such events are likely comparable to or longer than the current age of the universe, so are not likely to have threatened planetary habitability yet. However, as the universe ages, and the formation of new planets becomes rarer, sooner or later life will face this challenge. Will our planet's first extraterrestrial visitors be émigrés from a drowning world? Will they be mermaids? Life cannot last forever, but with technology it could last tens of billions of years, eventually moving out into the cosmos as the world that spawned it passes.

Conclusions

Gliese 581d and 667Cc are two close planetary worlds – lying within 25 light years of Earth. Gliese 581d takes the title as the first discovered potentially habitable world, while Gliese 667Cc is perhaps the most habitable world found, at the time of this writing. That said we shouldn't jump in the nearest large spaceship and head off just yet. Both worlds have potential problems when it comes to habitability for life forms like us. Gliese 581d appears fairly aged and most likely lacks a solid surface on which to land. It is also cold. Conversely, Gliese 667Cc is probably too young to have formed a breathable atmosphere and may, like Gliese 581d, have a surface dominated by water.

We may find that all planets with masses greater than two or three times that of the Earth are water-worlds. We may need to focus our attentions on the smaller, Earth-mass planets, where water is less abundant and a solid surface more probable. However, if Jack Lissauer's research is correct, and the habitable worlds of red dwarfs are deficient in volatiles, then we may not need to worry so much. For the reasons previously discussed (Chaps. 6 and 7), dry planets are likely to be more habitable for complex life forms than water-worlds; they will certainly be the likeliest abodes for sophisticated, fashion-conscious aliens, capable of space travel and communication. It seems adversity might adeptly be turned into opportunity.

These problems aside, the worlds found so far form the first of what will inevitably be countless potentially habitable planets. Should the ESA's Darwin, or NASA's Terrestrial Planetary Finder missions get off the ground, we should expect the discovery of a rich tapestry of habitable worlds within a decade. Sooner or later an Earth-sized world, of suitable age, will be found, and a second home for future humanity will emerge. However, there are many more variables to consider as to habitability than the mere location of the planet – mass, composition and age are all factors. Although tidal locking is unlikely to preclude the rise of life, the consequent failure of any geodynamo needs to be considered if the long-term habitability of the world is to be assessed.

These are early days of both discovery and understanding, and the brutality of hard data hasn't yet impacted on our imaginations. Although more hard work is due, there is still an abundance of opportunity for daydreaming.

Final Thoughts

Dwarfs, Metals and the Fate of Life in the Distant Future

Given that we know that red dwarfs will outlive all other stars in the universe, we can now turn our attention to the future of the stelliferous (star-forming) universe and the ultimate fate of life.

A little over 100 billion years from now, with our universe apparently set to expand forever, the vast majority of the remaining stars will be M-dwarfs. After the Andromeda Galaxy (M31) collides with ours in a few billion years, the amount of available hydrogen left to make stars will be very small – most having been used up in the starburst accompanying the collision, or expelled into intergalactic space by our galaxy's supermassive black hole. Consequently, by 100 billion years hence, the majority of star formation would have ceased as the amount of available hydrogen dwindled. Any remaining orange K-dwarfs will be set to die as helium-burning red giants soon after this time. These will be the last stars to engage in any form of complex chemical alchemy and will leave a retinue of very low mass M- and L-dwarfs to chart their dim, unconventional, yet persistent fates.

With red dwarf stars, too faint to see with the naked eye, this future universe would be dark indeed. To a planetary observer, the galactic hub will be reduced to a smudge of orange-red, with the ability to resolve individual stars limited. Only the astronomers of the future, equipped with the largest telescopes, could hope to see the last remnants of the once glorious stelliferous era.

A trillion years hence, the night sky will be a dismal sight indeed. Once the one trillion year mark has passed an important transition occurs. Up until this point the decline in stellar mass is largely offset by an increase in stellar radiation as stars evolve into red giants. However, at this advanced age only the smallest

D.S. Stevenson, *Under a Crimson Sun: Prospects for Life in a Red Dwarf System*, Astronomers' Universe, DOI 10.1007/978-1-4614-8133-1,
© Springer Science+Business Media New York 2013

stars remain, and these don't become such behemoths. Instead, as these diminutive objects evolve away from the main sequence they merely heat up (Chap. 3). Thus the galaxy begins a protracted phase of fading. At some stage, perhaps 10–20 trillion years in the futire, the last conventional star will wink out, leaving a frass of black dwarf stars.

The Last Stars Exit Stage Right

This seemingly depressing fate may not (quite) be the end of star formation and thus life. Not quite. Modeling work by Greg Laughlin, Fred Adams and others suggested that around 10–50 red dwarfs will be present in the galaxy, at any one time, for nearly a billion trillion years. These isolated, secondary red dwarfs will form as pre-existing brown dwarf stars sporadically collide and heat up to the point of hydrogen ignition. Each of these new red dwarfs will orbit the galactic center in among a sea of dead stars – mostly cold, helium-rich white dwarfs. Where a brown dwarf meets a dead white dwarf, more spectacular events can ensue, but these will be correspondingly rare. Thus an intelligent observer, stranded on a world orbiting a secondary red dwarf, will see nothing in the blackness of night. The overall luminosity of the entire galaxy will approximate the present Sun.

With the prospect of this future stygian gloom in mind, how, then, will the current stelliferous era end? Around a decade ago Greg Laughlin suggested a plausible finale. Currently, the lowest mass red dwarf has a mass approximately 1/13 that of the Sun, and has a surface temperature of only 2,200 K. These stars glow a dull red, with an atmosphere rich in partially ionized metal oxides and carbon monoxide. These stars are expected to endure for 12 trillion years.

However, a very curious effect occurs as the proportion of heavy elements (metals) increases. The minimum mass with which the star can ignite and sustain the burning of its hydrogen fuel decreases. Thus as the amount of metals increases the domain of the red dwarf will extend. Calculations suggest the minimum mass for hydrogen ignition could fall as low as 55 Jupiter masses.

Such a "star" would be odd indeed. The proportion of metals would need to be far higher than presently found in any star – exceeding 20 % by mass. This great mass of heavy elements would readily absorb energy from the slowly cooking core of the star, so much that very little would escape to the outside universe. With a core reacting at 2–3 million degrees and a diameter approximating that of Jupiter, you might imagine that the stellar photosphere would still broil at a 1,000° or more. However, such metal-rich stars would have atmospheres so rich in chemical compounds that very little of the core's output would be left to escape into space. Instead of a hot surface, the photosphere of these objects would in fact be utterly frigid by our standards. These stars would have less a photosphere as a thermosphere: a cloudy surface through which heat escaped. Temperatures at this "surface" would struggle to exceed the freezing point of water. To an outside observer the only output would be a drizzle of long-wavelength heat energy.

Such an object couldn't hope to support a biosphere on any neighboring world. Instead any living organism on such a world would have to draw upon the internal reserves of its planet. The star would be visible through an infrared telescope, but that would be all.

If these peculiar stars come into existence in the future, how might life adapt to this truly peculiar and frigid end to the universe's stelliferous era? Imagine the sight. Set in a pitch-black sky lies a star visible only in the infrared. In its core, hidden from view by clouds of water, ammonia-ice crystals, methane and complex organics, nuclear reactions simmer gently away. Would life adapt? Even with their main sequence life spans truncated by a metal content more than 10 times that seen in present day, they would still endure for trillions of years. Given the enormity of this life span, life – biological life – would have plenty of time to consider its options.

Conclusions

The future of the lowest mass stars is not yet cast in stone. Indeed even if one were to compare humanity's future with that of the Sun, many thousands of solar generations lie ahead before the feeblest red dwarf shows even an inkling of middle age spread setting in. Humanity will have to survive a long time to test these theories.

However, when the last true stars are formed, it is certain they will be red dwarfs. Whatever one's' political affiliations, it is inevitable that the reds will inherit the universe. The blues, by contrast, are but a mere chord in the universe's lengthy, but declining, repertoire.

After a long hiatus, red dwarf research is evolving fast. Finally, these lowliest of stars are finding their place in the astrophysical zoo as the future habitats for life. Stay tuned.

Glossary

Accretion   The process of the assembly of bodies such as planets and stars through the gravitational sweeping up of nearby debris.

Adenosine Triphosphate (ATP)   The molecule acting as the energy currency of all life on Earth. ATP consists of three phosphate ions linked end to end and attached to a 5-carbon sugar called ribose. The ribose is further attached to a base called adenine. Energy is stored in the bonds between each phosphate ion. ATP is also one of the four building blocks of RNA (ribonucleic acid).

Advection   The horizontal movement of energy from one area to another, such as the flow of warm air from sunlit equatorial regions to areas in darkness.

Amino Acid   An organic molecule consisting of three functional parts: an acidic section and a portion based on ammonia, linked through a third variable region. Twenty amino acids are used in the majority of life forms on our planet, and they are all left-handed.

Ammonia   A simple molecule made of one nitrogen atom linked to three hydrogen atoms. Ammonia is abundant in the atmospheres of gas giant planets in our Solar System and in the molecular clouds that spawn planetary systems.

Amphible   A silicate mineral containing magnesium and water as part of its structure.

Amphibolite   A rock produced by the alteration of basalt when it reacts with water. Amphibolite is an important driver of plate tectonics on Earth and powers the volcanoes found at subduction zones.

Angular Momentum   A property of all matter that moves in a curve or circle. Angular momentum, like linear momentum, is

D.S. Stevenson, *Under a Crimson Sun: Prospects for Life in a Red Dwarf System*, Astronomers' Universe, DOI 10.1007/978-1-4614-8133-1,
© Springer Science+Business Media New York 2013

always conserved, so when a planet moves in towards its star, other matter in its path must move outwards.

Aquaplanet   A planet whose surface is wholly or almost completely covered in a single, planet-wide ocean. Gliese 581d may well be an aquaplanet.

Atmospheric Collapse   The process by which an atmosphere loses some of its gases by freezing. On tidally locked planets the permanently dark side cools, and without a constant re-supply of warmth from the sunlit side will chill until first water and then carbon dioxide snow out. Except on very distant small worlds, any oxygen and nitrogen would remain at temperatures well above their condensing or freezing points.

Basalt   Quite possibly the most abundant volcanic extrusive (erupted) rock in the universe. Basalt is low in silicon dioxide (silica) and relatively rich in iron and magnesium. It consists of three minerals – olivine, pyroxene and plagioclase feldspar – of which the last is by far the most common.

Blue Hook (the)   The region on the Hertzsprung-Russell diagram linking the white dwarfs with metal-poor helium-burning stars on the horizontal branch. These are low mass stars that have lost most of their outer hydrogen-rich envelope and are burning helium in their core. They are at most a few times more luminous than the Sun, but very hot, with surface temperatures above 25,000 K.

Cannonball (Planet)   A planet with a disproportionately large iron core. Mercury is an example, with a core occupying 75 % of its volume. Other cannonballs may have proportionately even larger iron cores.

Carbon Dioxide   An important greenhouse gas found in the atmosphere of all terrestrial planets, bar Mercury. Carbon dioxide acts as an important thermostat in planetary atmospheres, since it can react with water and silicates to form carbonate (see above). Carbon dioxide is also the most important source of carbon in terrestrial photosynthesis.

Carbonate   A compound consisting of three oxygen atoms and one carbon atom. Metal carbonates, such as calcium carbonate, form an important group of rocks, including limestone, which stores carbon dioxide gas on our planet.

Carbonate-silicate Cycle   A series of reactions that occur on a planet involving carbon dioxide gas. Carbon dioxide in the air reacts with silicate rocks in a process called weathering. These reactions produce carbonates that wash into the oceans and are deposited. Geological processes can transport these into the mantle, where the carbonates decompose, releasing their store of carbon dioxide once more. This then is released into the atmosphere through volcanoes.

Carotenoid   Yellow, orange or red compound consisting of carbon, hydrogen and oxygen. Carotenoids absorb blue and green wavelengths of visible light, giving peppers and tomatoes their characteristic color. They also are part of the light-harvesting complex in chloroplasts and some bacteria that help these organisms carry out photosynthesis.

Centrifugal Force   An inertial and imaginary force caused by objects moving in curved paths. An object will tend to try and move in a straight line unless acted upon by a force (Newton's First Law of Motion). For example, a passenger traveling in a car rounds a bend and feels a force acting to push him or her outwards against the car. In reality this is just the effect of the force of the car engine pulling the car in a circle against the action of Newton's Law acting on the passenger, which tends to keep the rider moving in a straight line (experienced as inertia). Centrifugal force plays a part in the Coriolis effect.

Chemiosmosis   The complex mechanism by which bacteria, plant chloroplasts and the mitochondria use the energy of glucose to synthesize the molecule adenosine triphosphate (ATP). Hydrogen ions are pumped across a membrane, much like water building up behind a damn. These ions store energy in a chemical gradient that is then tapped into by an enzyme that manufactures ATP.

Chlorophyll (a and b)   Two related complex organic compounds that contain carbon, nitrogen, oxygen and hydrogen, arranged in four rings around a central ion of magnesium. Chlorophyll gives some bacteria and most plants their characteristic green color and captures energy for photosynthesis.

Chloroplast   A specialized structure, or organelle, in a plant cell that carries out photosynthesis. Chloroplasts have their own circular chromosome and are clearly derived from free-living bacteria called cyanobacteria through the process called endosymbiosis.

CN(O) Cycle   A series of reactions that occur at high temperature and fuse hydrogen to helium-4. The reactions get their name from the carbon, nitrogen and oxygen nuclei that are used to carry them out, serving as catalysts. Carbon-12 binds hydrogen and is converted into nitrogen-14, then oxygen-15, before this binds another hydrogen nucleus, breaking down to release helium-4 and the original carbon-12 nucleus.

Convection   The process by which heat is transferred from hot to cold regions by the movement of gases vertically. Once cool gases then descend once more and pick up more heat.

Coriolis Effect   A so-called inertial force experienced by objects (usually gases) that are moving across the surface of a rotating object. For example, gases near the equator of a planet that is rotating from west to east, then move towards the poles, will then be moving faster from west to east than the ground underneath them. This makes them appear to move eastwards.

Crust (Planetary)   The skin of mostly cold silicate rock that forms the surface of a planet. On most terrestrial planets this is made of a rock, such as basalt, that has been made by partial melting of the hot mantle underneath.

Cyanobacteria   A type of ancient bacterium that can carry out photosynthesis and produce oxygen as a byproduct. Cyanobacteria are often called blue-green algae and are the ancestors of the plant chloroplast that carries out the same function. Cyanobacteria may have appeared as early as 3.2 billion years ago.

Deoxyribonucleic Acid (DNA)   A complex molecule consisting of three main components wrapped up in a double-helix structure. The building blocks of DNA are a five-carbon sugar called deoxyribose, a phosphate ion and four carbon-nitrogen compounds – adenine (A), cytosine (C), guanine (G) and thymine

(T), also called bases. The precise order of the bases determines the genetic code that shapes the organism.

Differentiation (of Planets) Differentiation is the process by which planets become layered. When molten, or nearly so, heavier (denser) materials, such as iron, fall to the center of the planet, while lighter silicates rise above this. The lightest, or most volatile, materials, such as water, end up at the surface.

Eclogite A dense rock with a low silica content made up of two main minerals: garnet and pyroxene. Eclogite is produced by the metamorphism of basalt and is a key driver of plate tectonics.

Electron Transport Chain The name given to a chain of proteins and other molecules that transfer electrons from one molecule to another. In mitochondria, electron transport delivers electrons to oxygen gas and helps produce a chemical gradient needed to synthesize the energy molecule ATP.

Endosymbiosis The process by which bacteria and other single-celled organisms can invade a eukaryote cell (below) and take up residence. These invaders are then modified by complex processes until they become dependent on the eukaryote cell for survival. In turn the eukaryote cell benefits by gaining a source of energy or useful chemical compounds. The chloroplast of plants or mitochondrion of most eukaryotes is believed to have originated in this way between 1 and 2 billion years ago.

Eukaryote An organism made up of complex cells containing many sub-compartments that have specific functions. In most eukaryotes, for example, the mitochondrion produces most of the cell's energy in the form of ATP.

Feldspar A silicate mineral made up of silicon, oxygen and calcium, with varying amounts of aluminum, sodium and potassium. Feldspars are the most abundant minerals in most crustal rocks.

Flare A violent outburst that occurs on the surface of most low-mass stars. These have a complex origin but are related to the buildup of magnetic fields associated with sunspots. When the magnetic field lines become overloaded they snap, releasing their energy. Copious amounts of X-rays, ultraviolet light and ejections of mass can occur with each flare.

Fractional Crystallization   When a mixture of compounds in solution or when molten rock cools the compounds with the highest melting points freeze out first. These fall to the bottom of the liquid under the action of gravity and leave a less dense solution behind. This process can repeat many times until the resulting solution has a composition very different from the one that started.

Fusion (Nuclear)   The combining of lighter nuclei at high temperatures and pressures to make heavier nuclei. In the universe, hydrogen fusion produces helium-4 in the majority of stars and powers them.

Garnet   A colorful and dense aluminum silicate found in rocks formed at depth. On Earth this usually means at depths in excess of 35 km.

Gas Giant (Planet)   A massive planet in which the majority of the mass is contained as hydrogen and helium, often under such high pressure that it is both liquefied and turned into a metallic structure.

Granite   The foundation of all Earthly continents. A rock made up mostly of feldspar but with abundant quartz (pure silicon dioxide) and other iron-rich minerals such as mica. Granites have a low density compared with other igneous rocks on Earth, so they float over them. Granites are only produced in the presence of water and can be taken as a marker for the presence of past oceans on Venus.

Greenhouse Gas   A gas that is able to absorb radiation in the infrared portion of the spectrum and temporarily store it before then releasing it at longer wavelengths. In planetary atmospheres water vapor, carbon dioxide and methane are the most important.

Helium Flash   A violent reaction occurring in the core of some stars in which helium has reached its ignition temperature but is in a so-called degenerate state. Here, the particles are held rigidly together and cannot move freely when their energy increases.

Under such conditions helium fusion becomes explosive, and the sudden increase in temperature causes the degenerate state to lift. All low mass stars, including the Sun, will experience this at the climax of their red giant stage.

Hertzsprung-Russell (HR) Diagram   The eponymous diagram that compares the brightness of a star to its temperature or color. Developed near the turn of the last century by Henry Norris Russell and Enjar Hertzsprung.

Horizontal Branch (of Hertzsprung-Russell Diagram)   The strip of stars extending across the HR diagram at a luminosity roughly 100 times that of the Sun. Stars here have surface temperatures from 30,000 K down to around 4,000 K.

Hot Jupiter   A Jupiter-mass planet in a tight orbit around its star. Such planets are likely to have been born further out and then migrated inwards to their current position. All are expected to be tidally locked to their host star, always presenting the same face to the star as they orbit it.

Hot Spot (Geology)   A plume of hot mantle rock that rises up from near the core-mantle boundary. When these arrive near the surface much of the rock melts, producing large quantities of basalt magma. A long-lasting plume may create chains of islands as the overlying crust moves over it. The Hawaiian chain is one example.

Ice Giant (Planet)   A planet with a mass several times that of Earth, but dominated by water and other icy materials rather than gases such as hydrogen and helium. Uranus and Neptune are two nearby examples.

Ice IV   A state of water found at high pressures. This ice may be found near the bottom of the oceans of water-worlds or inside the moons of Jupiter's satellites Ganymede and Callisto.

Instability Strip (of Hertzsprung-Russell Diagram)   A band of stars with a temperature slightly hotter than the Sun but extending from the dim white dwarfs up to the very luminous supergiants. Stars in this region of the HR diagram pulsate as helium near the star's surface alternately gains and loses electrons, trapping and then releasing energy in waves.

Ion/Ionization   When an atom with no overall electric charge gains an electron it becomes negatively charged. When an atom loses an electron it becomes positively charged. The process of gaining or losing electrons is called ionization, and the term given to an atom that has gained or lost electrons is an ion.

K-class Star   An orange star with a surface temperature in the range 3,500–4,500 K. Some K-stars are giants, while others, such as Gliese 667A and B, are low mass dwarfs.

Komatiite   A silicate lava rich in iron and magnesium but poor in silica. Komatiites melt at high temperatures and were fairly abundant on the younger, hotter Earth and may be commonly erupted on Jupiter's moon Io.

Late Heavy Bombardment   A period stretching from 4.2 to 3.9 billion years ago when the terrestrial planets and our Moon suffered frequent catastrophic impacts from very large asteroids and comets. These may have originated far out in the Solar System when the outer giant planets underwent a period of migration to their current orbits. Such migration displaced a large number of icy planetesimals inwards towards the Sun.

Lithosphere   The outer rigid part of a rocky (terrestrial) planet consisting of the crust and topmost layer of the mantle.

Magnetosphere   A tear-shaped magnetized region surrounding planets with an active, circulating core. On Earth the magnetosphere is generated in the liquid outer iron core, but in giant planets it is generated either in a layer of circulating icy material or in metallic hydrogen – a form of very compressed and conducting gas. The magnetosphere shields a planet from the erosive effects of its star's wind.

Main Sequence   A diagonal strip on the HR diagram extending from faint red dwarf stars to bright blue giant stars. These stars obtain their energy by fusing hydrogen to helium in their cores.

Mantle   Usually the thickest layer in a terrestrial planet made up mostly of iron and magnesium-rich silicate minerals.

M-class Star   A cool star with a surface temperature between 3,500 and 2,200 K.

Metallicity (of Stars)   A measure of the abundance of elements heavier than hydrogen and helium found in stars. The oldest stars in the universe have a low metallicity (known as Population II) while stars such as the Sun are more metal-rich (Population I). Crudely, the metallicity of a star can then be used to estimate its age.

Metamorphic Rock   A type of rock that has been altered by heat and/or pressure. Eclogite is metamorphosed basalt.

Methane   A simple chemical known as a hydrocarbon and made up of one atom of carbon bonded to four atoms of hydrogen. Methane is an important greenhouse gas that can be produced by living organisms that live in the absence of oxygen.

Mid-ocean Ridge   A mountainous belt found often near the centers of ocean basins where new oceanic crust is being produced. Here the mantle partly melts under the axis of the ridge where the rock is under lower pressure, producing basalts.

Mitochondrion (Singular)   A structure found in most eukaryote cells that produces the energy molecule ATP in the presence of oxygen. Mitochondria (plural) have their own chromosomes and manufacture some of their own proteins. They are thought to have once been free-living bacterial cells that were acquired by earlier eukaryotes. The product of endosymbiosis.

Nice Group   A group of researchers based in Nice, southern France, which proposed that the outer planets weren't born in their current orbits but rather migrated there during the first few hundred million years of the Solar System's existence. The migration may, among other things, have been responsible for the Late Heavy Bombardment.

Nucleosynthesis   The name given to the series of nuclear reactions that creates all the elements in the universe. The Big Bang created hydrogen, helium and lithium, while nuclear reactions in stars subsequently produced all the other elements in the Periodic Table.

Olivine   A dense green iron or magnesium-rich silicate found in basalts, but making up most of the upper mantle of Earth, and presumably other Earth-like and super-terran planets.

Opacity   The ability of a substance to transmit light, or let light through. Substances that are opaque block the transmission of radiation; the wavelengths blocked depend on the type and physical state of the material. In a star, if the opacity is high energy it must be transmitted by convection or conduction. If it is low, radiation usually suffices.

Orbital Migration (Planets)   The processes through which planets may move during or after their formation in response to interaction with other debris in their vicinity. There is ample evidence for orbital migration in the early Solar System and also in many extrasolar systems where giant planets are found uncomfortably close to their parent stars.

Oxygenic Photosynthesis   The process of oxygen production through the capture of light energy in photosynthesis. On Earth, this process is confined to plants and several species of bacteria, such as cyanobacteria.

Partial Melting   A process whereby a mixture of compounds in a rock is separated by selective melting when the rock approaches its melting point. In the mantle rock near its melting point, partially melts. The densest mineral, olivine, with the highest melting point is largely left behind while minerals with lower melting points become liquid. Thus mantle peridotite partly melts to release basalt that has a lower melting point and is less dense.

Peridotite   A dense, green mantle rock predominantly made up of the mineral olivine. Lesser amounts of pyroxene and feldspar make up the remainder of the rock.

Perovskite   Under high pressure iron and magnesium-rich minerals (such as olivine) in peridotite are transformed into a denser mineral called perovskite. On Earth, the lower mantle is made of perovskite while the upper mantle is dominated by olivine.

Photic Window   The depth of water in which the amount of available light is sufficient to power photosynthesis.

Photosynthesis   A series of chemical reactions that occur in some species of bacteria and the chloroplasts of plants. Light is captured in one series of reactions that produce oxygen, the energy molecule ATP and another molecule that carries hydrogen

(NADP). In the second series of reactions NADP and ATP are used to convert carbon dioxide into glucose.

Photosynthetic Active Radiation (PAR) The wavelengths of light that can be captured by an organism to drive the process of photosynthesis.

Plagioclase Feldspar A silicate mineral containing the elements calcium, oxygen and silicon.

Planetary Nebula The final stage in the active life of a low mass star like the Sun. The core contracts to become a hot white dwarf, which releases copious amounts of ultraviolet radiation. The ultraviolet radiation causes the expanding and dispersing hydrogen and helium-rich envelope of the dying star to fluoresce. The shell of fluorescing gas is called a planetary nebula.

Planetesimal A roughly circular body of ice and/or rock a few hundred kilometers across that are formed from the collision of smaller bodies of material in orbit around young stars. Planetesimals may collide over hundreds of thousands of years to produce larger, planetary bodies.

Plate Tectonics The process through which the surface of Earth is broken into a number of independently moving pieces. Plate motion is driven by convection within the hot, mobile, underlying mantle.

Population I Star A star with a relatively high content of metals (elements heavier than helium) that is similar to or greater than that of the Sun. Population I stars are typically less than 8–9 billion years in age and belong to the disc of stars orbiting in the mid-plane of the galaxy.

Population II Star Ancient star with an orbit that is often highly inclined to the disc of the Milky Way. Population II stars have metal contents less than one hundredth that of the Sun and have ages extending back to a few hundred million years after the Big Bang. Many are organized into tight groups of stars called globular clusters.

Pressure Gradient (of Air) A difference in the pressure of air in the atmosphere of a planet. Areas of rising air have lower pressure than areas where air is descending. The difference in pressure causes air to flow from areas of high to low surface pressure along this gradient.

Protein   A complex molecule made up of tens to hundreds of copies of smaller subunits called amino acids. In most organisms on Earth there are 20 common amino acids used in every protein, with the uniqueness of the protein defined by the order of the amino acids that are used.

Proton-Proton (pp-) Chain   A series of connected reactions in which hydrogen is fused in fours to produce helium-4. There are different variants of the process depending on the temperatures present in the core of the stars running them.

Pyroxene   A silicate mineral containing the elements silicon, oxygen and magnesium. Pyroxene is a common mineral in Earth's mantle and in the basalts that are abundant on Earth, the Moon and other terrestrial planets.

Quartz   The simplest silicate consisting of silicon and a pair of oxygen atoms in the formula SiO_2.

Recombination (of Electrons)   In some gases the atoms of different elements have gained or lost electrons and become ionized. Where an electron has been lost, the electron will have gained energy from its surroundings. It can then lose this energy once more and fall back onto the ion that was created when it left. The return process, where an ion and electron come together, is called recombination. Recombination is an important process in stars with surface temperatures marginally higher than the Sun. Alternating waves of ionization and recombination cause these stars to pulsate. See RR Lyrae Variable.

Red Dwarf   An abundant, small, dim class of star with a mass of between 0.55 and 0.075 times that of the Sun. Red dwarfs make up 75 % of the total number of stars in the visible universe, with over 150 billion in the Milky Way.

Red Giant   A star with a mass similar to or somewhat greater than the Sun that has exhausted its central store of hydrogen fuel. Its core now contracted to the size of Earth, the outer layers expand and cool down. Red giant stars currently measure up to a few hundred times the diameter of the Sun, but future giants will be proportionately smaller as the stars that give rise to them dwindle in mass.

Respiration   Not breathing, but the process by which organisms generate usable energy in the form of the energy molecule ATP. Respiration can be either anaerobic (occurring without oxygen) or aerobic (requiring oxygen). Aerobic respiration is far more efficient but is confined to some species of bacteria and organisms that contain mitochondria.

Ribonucleic Acid (RNA)   A type of long polymer molecule used to store or transport genetic information. Some RNA molecules can act as biological catalysts, carrying out essential functions within the cell, such as the formation, or synthesis, of proteins.

RNA World (the)   A hypothetical era in Earth's biological history where RNA served as both the cells' store of genetic information and acted as catalysts in lieu of DNA and proteins, respectively. The RNA World has good evidential support in the form of various ancient cellular molecules that require RNA in a functional or support role. The manufacture of proteins involves several species of RNA catalyst, including the critical ribosome where amino acids are combined in sequence to manufacture proteins.

RR Lyrae Variable   A type of evolved giant star with a surface temperature in the range of 9–11,500 K. These stars have diameters not much greater than the Sun, but have a luminosity 100 times as great. They pulsate over the period of a few hours to a day as helium, lying in a narrow region underneath the stellar surface, alternately heats and ionizes (loses electrons) or cools and recombines with these electrons. Most RR Lyrae stars belong to Population II.

Serpentine   A dense silicate mineral rich in iron and magnesium that has been thoroughly altered through interactions with hot water. Serpentine began its life as the mineral olivine in the heart of freshly minted oceanic crust. Seawater percolating through the hot crust reacts with the olivine, generating serpentine and hydrogen gas.

Serpentinite A rock with a composition dominated by the mineral *serpentine*.

Silicate A mineral containing the elements silicon and oxygen, with varying amounts of iron, magnesium, calcium, aluminum, sodium or potassium. Silicates make up the vast bulk of oxygen-rich planets like Earth, Venus and Mars.

Star Spot A sunspot on another star. Star spots contain localized strong magnetic fields that restrict the flow of heat from the star's interior. This makes them hundreds of Kelvin cooler than the surrounding stellar surface (photosphere). The magnetic fields of star spots often spawn flares and other stellar eruptions that can have significant effects on the atmospheres of neighboring planets.

Stellar Climatic Habitable Zone (SCHZ or Stellar Habitable Zone) The region around a star where the amount of radiation arriving at the surface of a planet would raise its temperature to within the freezing and boiling points of liquid water. Stellar climatic habitable zones are narrow around low mass stars but progressively widen as the central star is increased in mass.

Stromatolites Rocky formations built by communities of bacteria growing in the shallows of warm seas around the globe. These communities are organized around the photosynthesizing hub created by cyanobacteria. Stromatolites are among the most ancient fossilized structures found on Earth.

Subduction A process by which cold, dense oceanic crust sinks back into the hot underlying mantle. Subduction is a key driver of plate motion (plate tectonics) on Earth and perhaps other rocky planets. Subduction recycles carbon dioxide, maintaining habitable conditions on Earth. It also drives the formation of most continental crust, which may be necessary for the genesis of complex life.

Sub-stellar Point (SSP) An imagery point on the surface of a tidally locked planet where the light from the central star shines directly down from above. All habitable worlds orbiting red dwarfs will be tidally locked and thus have a SSP.

Super-Rotation An interesting and unexpected pattern of wind flow on planets that are either slowly rotating relative to their orbit or are tidally locked to their star. Winds blow in a direction counter to the rotation of the planet, with this pattern of airflow primarily confined to equatorial regions.

Super-Terran (Super-Earth) A planet that is primarily rocky and with a mass in the region 2–10 times the mass of Earth. At higher masses these planets will steadily take on the appearance of Uranus and Neptune as proportionately greater masses of gases (volatile elements) are captured by the planet's increasing gravity.

Thick Lid A term given to the formation of a stagnant upper (or sometimes lower) layer in a planet that resists further convection or deformation driven by convection within an otherwise hot planet. In the context of super-terrans the most likely scenario of their evolution involves the formation of a thick layer of buoyant granite at the planet's surface. The low density of granite prevents it from subducting, causing the process of plate tectonics to fail.

Tidal Locking Any planet orbiting at a suitably close distance to its parent star has tides raised within its atmosphere, oceans and mantle caused by the star's gravitational pull. This effectively breaks the planets spin, causing its orbit to circularize and change in distance from the star. Tidal locking in the Earth-Moon system caused one face of the Moon to stare permanently at Earth; Earth's rotation to slow and the distance between Earth and the Moon to steadily increase.

Transcription (RNA) A series of chemical reactions that occur in cells that allows them to copy the genetic information in DNA to a messenger RNA molecule. Typically, but not exclusively, transcription generates either an RNA molecule that codes for a particular protein or produces an RNA that has a directly functional role in the cell.

Translation (Protein) Messenger RNA (mRNA) molecules contain a series of instructions for the correct assembly of amino acids into proteins. Translation refers to the change in language from genetic code to amino acid code and the process of decoding is carried out by the cell's ribosomes.

Water   A very simple, abundant molecule made up of two parts hydrogen to one part oxygen (H_2O). Water is unique in its properties – a propensity to dissolve a very wide range of substances, melting and boiling points that are conducive to the operation of carbon-based life and an ability to hold itself together through weak bonds that in turn allow it to flow as proficiently across landscapes as it does through the interior of every life form on Earth. Water may be a prerequisite for life.

White Dwarf   A dead star with masses currently between 1.37 and 0.2 times that of the Sun. These stellar corpses have an inert, degenerate core made of either oxygen, neon and magnesium; carbon and oxygen; or, at the lowest masses, helium. Aside from destructive pairing with more conventional stars, the fate of all white dwarfs is to slowly cool down until they are as cold and as desolate as surrounding space. All stars with masses less than ~8 times that of the Sun will end their lives this way. Cooling is initially fast, but steadily slows as the temperatures fall. After nearly 250 billion years of cooling a typical white dwarf will still be hotter than a pan of boiling water.

Index

D.S. Stevenson, *Under a Crimson Sun: Prospects for Life in a Red Dwarf System*, Astronomers' Universe, DOI 10.1007/978-1-4614-8133-1, © Springer Science+Business Media New York 2013

CPSIA information can be obtained at www.ICGtesting.com
Printed in the USA
LVOW02s1455131015

458074LV00002B/5/P